MW00826475

The Successful Race Car Driver:

A Career Development Handbook

Other SAE books on this topic:

The Golden Age of the American Racing Car
By Griffith Borgeson
(Order No. R-196)

Race Car Vehicle Dynamics
By William F. Milliken and Douglas L. Milliken
(Order No. R-146)

For more information or to order this book, contact SAE at 400 Commonwealth Drive, Warrendale, PA 15096-0001; (724)776-4970; fax (724)776-0790; e-mail: publications @sae.org; web site: www.sae.org/BOOKSTORE.

The Successful Race Car Driver:

A Career Development Handbook

Robert Metcalf

Society of Automotive Engineers, Inc.
Warrendale, Pa.

Library of Congress Cataloging-in-Publication Data

Metcalf, Robert (Robert E.)
The successful race car driver : a career development handbook/
Robert Metcalf.
p. cm.
Includes bibliographical references and index.
ISBN 0-7680-0497-7
[1. Automobile racing drivers—Handbooks, manuals, etc.
2. Automobile racing—Vocational guidance—Handbooks, manuals, etc.]
I. Title.

GV1029 .M49 2000

99-045955

400 Commonwealth Drive
Warrendale, PA 15096-0001 U.S.A.
Phone: (724)776-4841
Fax: (724)776-5760
E-mail: publications@sae.org
http://www.sae.org

ISBN 0-7680-0497-7

SAE Order No. R-254

Dedication

For my father, Robert P. Metcalf, who taught me that anything is possible if you want it enough

Contents

Part Two: The Four Things That Every Successful Driver Learns

Part Three: Going Pro

Author's Note

As we enter the third millennium, life is very easy for many of us. We live in a world of satellite communications, microwave ovens, and laptop computers of amazing power. However, along with these modern conveniences have come modern problems. Lawyers chase ambulances, hospitals overcharge patients because insurance companies will pay their bills, and the government's antics are enough to make Thomas Jefferson and Ben Franklin turn over in their graves. At each opportunity, individuals in our society try to pass the buck.

Americans like to brag that their country has "the most freedom" of any country in the world, but as problems have sprung from our conveniences, so responsibilities follow those freedoms. As Viktor Frankl once commented, we must supplement the Statue of Liberty on the East Coast with the Statue of Responsibility on the West. In our current society, many individuals do not abide by this philosophy and can (and many times, do) sue one another for any reason. Right or wrong no longer seems to be a part of that decision. Because of this deplorable situation, I have been advised by my attorney to include the following disclaimer—even if I only give you advice on opening a soft drink.

My firm belief, and indeed that of our legal system as detailed in the constitution, is that each of us is responsible for our own actions. Any activity in which man is in motion and the object is to push the performance envelope is inherently dangerous, and if you knowingly participate in a dangerous activity, you have made a conscious, rational decision to do it. The techniques explained in this book are intended solely for use in controlled competition events, not on public roads. They have been used successfully many times before by many drivers in such events. Although some of us have fallen off the road or experienced other calamities from time to time, these accidents have been a result of our own decisions and actions. Analyze the advice in this book to see if it is applicable to your own situation. If it is not, do not use it. If you do choose to adopt any of the techniques, tips, attitudes, procedures, or other advice detailed in this book and an accident happens to you, realize that it is a result of your own conscious, rational decisions and actions. I disavow any and all responsibility for your accident and any damage or injury resulting from it. Racing is a dangerous sport. If you freely choose to participate in racing and use advice obtained from this book, accept that you are doing something dangerous and accept responsibility for your decisions and actions.

Prologue

Why Do We Race?

Ernest Hemingway was one of the world's great individualists. There have been many others. Vincent Van Gogh, John Lennon, Albert Einstein, and Henry David Thoreau were not easily led by others, but it was Hemingway who said, "There are only three real sports in the world: bull fighting, mountain climbing, and auto racing. All others are just games." You know that or you would not have picked up this book. Football, baseball, soccer, and all the rest of the stick-and-ball sports have their place, and I'll not try to diminish the skill and dedication it takes to excel in any of these. Few are able to compete with Troy Aikman, Michael Jordan, and the others involved in professional sports. They are supremely fit athletes and have honed their quite considerable skills through years of practice, patience, and dedication.

If you make it to the top levels of auto racing, you will have developed the same levels of skill through the same years of practice and patience and the same level of dedication, but with the added element of risk. On the rare occasions when Tiger Woods makes a really big mistake, his ultimate penalty is his ball landing in a bunker. It's embarrassing, frustrating, and increases his score. I have personally watched drivers make very small mistakes on racetracks and pay the ultimate price, their lives. This doesn't happen often, but it does happen, and it's not only in the top levels of motorsports. It can happen just as easily in a Formula Ford as in an F/1 car. It is that element of risk that separates the "games" from the "sports." When everything goes wrong, you will wonder why you ever got into the bloody sport in the first place, but there is nothing else like it when everything is right.

Some would consider just driving a race car at speed to be a fate worse than death. Consider this: You have strapped yourself into a quivering, featherlight machine, surrounded by 40 gallons of flammable racing fuel, with an engine developing 850 horsepower inches from your back. You are traveling at three and a half times the highway speed limit, and at that speed the small ripples and bumps in the pavement almost jar the steering wheel out of your hands. Your body weight soars to almost four times normal because of the G forces, but in odd directions and sometimes in two directions at once. The hundred and fifteen decibels of noise deafens you and the vibration of the 13,000 RPM engine renders you almost blind. The oil on your visor and the sun in your eyes finishes that job. But that only sets the stage; now comes the test. As you approach a turn, you must brake as hard as possible without locking a wheel and losing precious grip, downshift two gears, turn the car with high G force level, and place it with precision to within a few inches of a spot on the pavement, which you used as a reference on the previous lap, hoping nothing has changed since you entered the same turn two minutes earlier. All the while, you are defending your position from the two other cars who are both trying to steal the piece of pavement you need going into the turn. This scenario is played out turn after turn, lap after lap, for two or three hours during many races. Demands are placed on the racing driver that are seldom even suggested elsewhere.

The challenge of meeting those demands, and the penalties if you don't, make auto racing a haven for the individualist. Racing is an art form. It demands creativity. It is a discipline and demands that as well. It requires a very high level of intelligence, concentration, and tough-mindedness. If these qualities are not already part of the race driver's makeup, they can and will be developed as he gains experience. As other people have found their own havens in music, literature, and art, those of us who race find ours in the greatest of the true sports, auto racing.

Acknowledgments

In the past, whenever I picked up a new book, I was often amazed that the author had so many people to thank. Now that I have been on the other side of the pen, however, I understand why. Just as the winning driver on the podium does not get there by himself, an author does not turn an idea into a book without considerable help. Writing a book and getting it published is a team activity. Although I have done the writing myself, many other people have helped to produce what you now hold in your hands. So here goes. Just picture me in a driver's suit on a podium thanking those responsible for the win …

First, I want to thank all those readers of *Sports Car* magazine who read my articles and gave me encouraging comments when we bumped into each other at race tracks across the country. Although I had much of the book written when those articles were published, the kind words kept me enthusiastic about getting the book out. Thanks, also, to my editor at *Sports Car*, Rich McCormack. When the majority of the book was completed, several of my friends read the manuscript and made various suggestions about the ideas contained in it and the wording used, and occasionally pointed out typos and misspellings. Although too many of them to list have helped in this way, I thank them all, but particularly Bob Norwood for his words of encouragement. Mark Weber provided many of the excellent photos that illustrate some of the ideas found in these pages, and Paul Laguette did an excellent job on the illustrations. Bill Brown of Road Atlanta, Karl Uberbacher of the Skip Barber Racing School, and John Dambros of Simpson Products also provided photos or illustrations. Mike Stephens of Stephens Brothers Racing School went to a lot of trouble to make an excellent series of photographs of "the Bitch" at the Hallett Motor Racing Circuit. The book may not have been published without Bill Milliken, author of *Race Car Vehicle Dynamics,* who put me in contact with the publisher, the Society of Automotive Engineers.

Most of all, though, I would like to thank all of the drivers whom I have had the opportunity to work with or just observe over the years. They are the people who have taught me all of their hard-learned lessons, many times by good example, but sometimes by sharing with me how not to approach a problem, a lesson that is just as important. I have learned from many, but the ones from whom I have learned the most include Craig Taylor, Brian Goellnicht, Price Cobb, and Michael Roe. Thanks, guys!

Introduction

The race cars of the late '50s did not closely resemble those we see on today's tracks. Back then cars were larger, heavier, much less nimble, and would not corner or accelerate nearly as fast as today's race cars. Few knew anything at all about engine power bands or oversteer/understeer balance. There were few sponsors and little need for them. Electronics in racing was unimagined. Only hard, skinny tires were used and drivers had little safety equipment. Auto racing has changed a great deal since then. Carburetors have been replaced by electronic injection, computers now control engine functions, shock absorbers cost $5,000 a set and their adjustments are some of the most important on the car. Wheels are sometimes wider than they are tall, and steel tubing frames have given way to those made from carbon fiber fabric and epoxy. Race cars have become much more refined and require sophisticated techniques to prepare them and to drive them. Indeed, the entire racing industry is more sophisticated, requiring a professional approach to obtain successful results regardless of the level of competition.

Racing today is characterized by high technology—technology requiring the knowledge and experience of individuals trained in its use. No successful team can now afford to be without a competent engineer and large teams break this job down further, designating specialists for tires, shocks, data acquisition, data analysis, engine management, and other tasks. Teams now have composite technicians, brake specialists, power transmission specialists, and suspension experts rather than the mechanics of previous decades.

To use the growing technology effectively, more complex methods and procedures have been developed, which are not only applied directly, but also come into play in overall racing functions. For instance, because tires have become stickier, and the adhesive qualities of the rubber are affected by heat, tire wear and tire hardening must be calculated in determining pit stop strategy. Even when pit stops are not required, the heat cycles experienced by the tires must be considered if the team is to be competitive and stay within a specified budget. Such considerations were not required only a short time ago.

All of the new technology, specialized knowledge, and new procedures cost money, and so, just as technology characterizes the sport, big business fuels it. Racers began to see in the late '60s just how valuable the sport could be to business, and as they took advantage of that value it spurred progress. Consequently, we now find both business and technology to be inseparable components of racing.

In motorsports, the boundaries separating technology and business from other fields are somewhat fuzzy, however. The business and technology aspects often carry over into driving, racing, and dealing with officials, spectators, business associates, and sponsors. The fact that racers are always looking for an advantage frequently puts them in a position of dealing with officials on both technical and political grounds. For example, the racer may be required to explain to a sponsor why a new bit of technology is required or why it costs so much. Then, those who will be using the new technology may need to be educated. Finally, the new technology may necessitate a change in the method of car preparation or in driving style. Thus, as with most areas of life, in racing, everything affects everything else.

Most club racers do an acceptable job of learning to drive and mixing it up with their competitors, and if we were living in the '60s, that would be enough. Now, however, even the amateur racer must be knowledgeable of the effects of business and technology on the sport and have some idea of how to use them to his advantage. If he is not, his racing is likely to produce less than acceptable results and be frustrating at best and dangerous at worst. If he has intentions of continuing with a racing career, inability or inattention in either of these areas will certainly limit his opportunities to proceed.

During the last four decades, power has sometimes been transferred correctly from drivers and mechanics to engineers and businessmen, but more often than not, this process seems to yield unacceptable results. This is not because racers lack intelligence. In fact, they are among the most intelligent people with whom I have come in contact. The problem seems to be that they either fail to realize the dynamic nature of auto racing or fail to learn enough from the racers who went before them and continually reinvent the wheel.

Trying to learn from other racers is not always a satisfactory approach. For example, when you ask another racer how to take a particular turn or how to set up a car for a particular track you will sometimes get an answer that is exactly opposite of what he really thinks. And even if the other racer is being completely honest and has a genuine desire to help, he may give inaccurate information. Thus, a more effective approach is to simply observe other racers. Because of the proliferation of misconception and misinformation, paddocks and race tracks provide great opportunities to see mistakes in the making. It is common to stand beside a track and see an incorrect line used when a driver runs out of road at the exit of a turn. Seeing a Formula car come in dragging its rear wing behind it or watching as a brand new racing engine is sawed in half because the driver was trying to cheat with nitromethane are things you will remember.

The more subtle mistakes have more serious and longer-lasting career consequences, though, and are usually made in the office or in the garage between races when no one else is around. Such mistakes inevitably involve attitudes or thought patterns and are thus reflected in the decisions the racer makes both on and off the track. In working with a number of quite successful drivers over the years, I have had the opportunity to learn the criteria on which they base their decisions and the attitudes they maintain in implementing them. These attitudes include their views on physical and mental fitness, car preparation, driving style, setup of the car, dealing with officials and sponsors, and a host of other issues.

Just as attitudes are important to a racer's success, proper driving and car setup skills are indispensable. Thirty years ago most drivers thought that taking the proper geometric line through a corner was the fastest way. Since then, tires have become stickier, suspensions have become more sophisticated, cars have sprouted wings, cornering speeds have risen drastically, and cars react much more precisely and with more force than in previous times. These changing response characteristics of the newer cars require that drivers learn new skills. Anyone can learn the skills required to drive a car fast. Combining those technical skills with the proper attitudes is a bit more difficult, and it is that combination that seems to separate the successful racers from those who feel frustration at each event.

People drive race cars for one of two basic reasons: for fun or as a professional career. Those who do it for fun generally have jobs and families and do it as a hobby. Even so, they can be quite serious about it, spending a great deal of time, effort, and money racing at tracks across the country for many years. Some of them win races regularly and are more successful than others who struggle on for years never really achieving the success they seek. Obviously, of those who do become professional drivers, some are more successful than others, too.

Regardless of your motives for wanting to be involved in auto racing, the path you must follow and the lessons you must learn along the way in today's high tech and big business world are the same. As Steve Cameron, Lynx Racing's Team Manager explains, "No matter how fast they (drivers) inherently are, there is a process they need to be guided through to turn them into true champions." You will find that process described in the pages of this book. You will also learn valuable skills and techniques known by all good drivers for driving and car preparation and for dealing with officials and sponsors, as well as tips and tricks to help you avoid pitfalls along the way. Some of these may keep you from making mistakes that would cost you thousands. Some will make you faster. All of them will help you to more fully enjoy the sport. If your interest in racing is such that you are content to remain an "innocent bystander," I recommend *The Successful Race Car Driver*. This book is full of insider information that will give you a better understanding of what Little Al, Rusty, and Jacques are going through and why they make some of the decisions they do.

Since its beginnings in the 1890s, auto racing has been dominated by men. In this book, male pronouns are used exclusively, but this should not be construed as excluding women from racing. Women, including Kara Hendrick, Deserie Wilson, Janet Guthrie, and Lyn St. James, all of whom have proven themselves to be extremely capable drivers, have competed at the highest levels of motorsports. Being male is not a prerequisite to being a competitive driver. It is, however, still a sport dominated by men, so 'he' will be used when referring to a driver rather than 'he/she.' We, as racers, so far at least, tend to be more concerned with performance than with political correctness.

Traditionally, road racers and oval track racers have not gotten along well. The oval track guys say, "Road racing is for the wine and cheese crowd," and the road racers respond with, "I don't want to run ovals, they're boring." It is time to finally put this rivalry to rest. Every competent racer today knows how to run both and does not harbor animosity for a particular type of racing. Winston Cup cars run Watkins Glen and Sears Point, two of the toughest road courses in the country. Indy cars run big ovals and small ones, as well as street circuits, road courses, and airport tracks. To succeed in modern motorsports, the racer should develop the attitude that the type of racing he does is immaterial, but adapting to the prevailing conditions is important, whether those conditions include left turns or right.

In the following pages, we will deal with both. As a matter of fact, the only real differences are in equipment and driving techniques. A racer must still deal with engineers and sponsors the same way, and the same organizational skills are required regardless of the type of track he races on. These days the only distinctions that can properly be made in racing are between what we will call "circuit racing" (which includes oval and road racing) and "drag racing," which admittedly is somewhat different. Even drag racers, though, share many qualities with circuit racers.

Racing is a very complex activity requiring the driver to be proficient at a number of skills. And, as with any skill, learning takes practice. As you gain experience in each aspect of motorsports, you may discover some techniques not described in this book which will work better for you. None of the rules of racing are cast in stone. (Even those in the sanctioning body's rule book may be somewhat flexible depending on who is involved.) Every successful driver develops his own techniques with experience on the track, and this is true of the off-track endeavors of racing, too. There is no one correct way to do it. The techniques described in the following pages are a good place to start. After you gain experience, don't be afraid to experiment and find out what works best for you. The goal is to find the *best* solution. The team that finds the best solutions to the day's problems wins the race.

Racing takes dedication, education, skill, and a little luck. Dedication is a quality you must build up within yourself. Education can be obtained through driving schools, books, tapes, and by listening (selectively) to other racers. Skill comes with practice and experience. Even luck is a matter of being in the right place at the right time. To put it another way, luck is a matter of who you know, what you know, and having a good sense of timing. This is a difficult sport to master, and it is becoming more difficult as the influence of business and technology increases. Although you may become proficient at it, you will never perfect it. You will never reach the end of the learning curve. It is doubtful that Andretti, Earnhardt, Schumacher, or any other experienced, top driver would say that he has mastered the sport or cannot do better. This book contains information that can help you make the proper decisions to work your way through the maze of people and activities that make up auto racing, with a minimum of frustration, while maximizing success, and as a result, enjoyment. Racing is always more fun when you win. I hope that, through these pages, your position on the learning curve will be advanced so as to increase your number of wins and your enjoyment of the sport.

Robert Metcalf
Dallas, Texas

Part One: Making a Commitment

Chapter 1

Mind and Body

In all the world of motorsports performance is the most important consideration—both mental and physical.—Mark Martin

Psychological Elements

Racing is life. Everything that happens before or after is just waiting.—Steve McQueen in *Le Mans*

To the layman a racing driver is a courageous daredevil with perfect eyesight and feline reflexes, and someone who harbors a death wish beneath a suave movie-star exterior. A real racing driver, however, is an average person with many qualities, both psychological and physical, in various proportions. Some drivers seem to have an innate gift, but having that gift is not a prerequisite to being a successful driver at any level. Instead, it is the particular combination of qualities that the driver possesses, and the degree to which those qualities are developed, that will determine the level he will reach and how he will choose to enjoy the sport.

Rarely in the history of motorsports has there been a driver who showed as much raw, innate talent as Gilles Villeneuve. Gilles began his racing career in snowmobiles in his native Canada, but quickly moved through Formula Ford and into Formula Atlantic. His remarkable drives at Trois Rivieres caught the attention of anyone fortunate enough to see him, including James Hunt, then driving for Team McLaren. During a short stint with McLaren, Gilles gained Enzo Ferrari's attention, and the next year moved to Italy. In the red cars from Maranello, Gilles quickly became one of the quickest and most flamboyant drivers in Formula One. His style behind the wheel could only be characterized as "flat out." His method of getting around the race track was to keep the throttle open for as long as humanly possible. Of course, this meant he was sideways a good portion of the time. It would have been very interesting, had we the technology then, to equip his car with a data acquisition system to monitor how long he stayed on the throttle into a turn, how soon he opened it up again at the exit, and what his corner exit speeds were in relation to other drivers of his time. His car control and ability to continue these antics lap after lap are now as legendary as his inability to temper his performances to accommodate existing conditions.

Although Gilles was able to keep the car pointing in the proper direction most of the time while keeping his foot on the throttle, he did make more than his share of mistakes. Gilles drove by the seat of his pants. He was *very* fast, but he still drove by the seat of his pants. His performances on the track were in sharp contrast to his shy, boyish manner in the paddock. He was calm, easy-going, and genuinely liked by everyone. Gilles was killed while on a qualifying lap for the Belgian Grand Prix in 1982.

It has been said that Michael Andretti's style in a race car is "aggressive." Some may remember his win at the 1994 Surfer's Paradise Indy Car race, the first of that season, in the maiden voyage of the Reynard. Early in its development, Adrian Reynard's new creation had an almost incurable problem with corner entry understeer. In the Australian race, Andretti's method of overcoming this difficulty was to change his line to drive over the curbs so that he would not have to turn as sharply and, therefore, not generate the understeer. This produced some spectacular high jumps. After the previous season of his dismal debut with McLaren in Formula One, his determination to show the world that he was still a world-class driver prompted him to put on this show and win the race.

After struggling to learn the Grand Prix ways for most of the previous season, Andretti's only podium finish was after a truly magnificent drive from 22nd position due to an early pit stop. He drove smoothly, and quickly, and he made clean, effective passes. He was not at all flamboyant. The TV commentators did not even know he was moving up until he appeared in the top ten. He continued to drive quickly and cleanly and pick off each competitor in turn until he found himself in third position hounding Jean Alessi on the last lap and looking for a way around. It is unfortunate that McLaren had already decided to release him from his contract at the conclusion of that race.

Andretti's coldly calculating and aggressive style contrasts sharply with Villeneuve's flat out technique. Few can manage Villeneuve's style for long without a mistake. The skills required to drive a race car can be learned. Each of these drivers had to learn the skills of car control as will any beginning driver. Similar to playing the guitar, using a computer, or skiing, any skill can be learned and perfected by learning specific techniques and practicing them. Sure car control skills are required to drive a car, but what does it take to drive one *well?* What sets the really good driver apart from those who never become successful? Whether Andretti's or Villeneuve's technique is better is less important than why each man drives as he does. The answers to these questions requires us to know a bit about how the racing driver is put together psychologically.

Everyone's personality is made up of a number of traits which psychologists can measure. These range from integrity to cynicism. The qualities that an individual possesses, and how and in what ratios they are combined make the difference. To determine which qualities are important to be successful, we should look at those traits that have actually been measured in successful drivers.

In the late 1960s, in San Francisco, amid free love, flower power, and Jefferson Airplane concerts, psychologist Dr. Keith Johnsgard conducted a survey of the psychological makeup of racing drivers. It all began when the San Francisco Region of Sports Car Club of America (SCCA) asked Johnsgard, who was known to haunt the paddocks of West Coast tracks, to produce a test that could be used to screen driving school applicants to determine which individuals would drop out of racing, which would continue, and which might present dangerous situations on the track. Dr. Johnsgard soon enlisted the help of an East Coast colleague, Dr. Bruce Ogilvie, who had a great deal of experience with such psychological testing of both amateur and professional athletes, having interviewed more than 4,000 of them. Dr. Ogilvie had written a book on the subject and was better informed on the psychology of athletes than anyone else in the world. The San Francisco Region's test developed by Johnsgard and Ogilvie soon became a profile of drivers of all levels of experience—from the regionally licensed club racer to the pros of Formula One. Some of the drivers tested 30 years ago included Graham Hill, Jackie Stewart, Mark Donohue, Jim Hall, Dennis Hulme, and many other top professionals. The results stand today as meaningful and insightful as they were then.

Meanwhile, at St. George's Hospital in London, Dr. Bernice Kirkler was assessing the likelihood that a driver who had suffered a severe head injury would return to the track. Since she regarded racing drivers as somewhat unique individuals, it was necessary for her to gauge their performance against that of ordinary British motorists to establish a standard. She asked five Grand Prix drivers to come to the hospital for a number of tests. Five lay drivers were also examined.

In addition to determining the injured driver's level of recovery, Dr. Kirkler found that under normal driving conditions the performance of the race drivers and the British motorists was almost identical. However, when speeds and stress levels were increased, the performance of the Formula One drivers increased while that of the average motorists declined sharply. Having heard of Dr. Kirkler's findings, Dr. Johnsgard said, "Regarding this unusual ability to meet exacting demands under conditions of great stress, our study shows that a combination of important traits is necessary: emotional stability, self control, and basic low tension or anxiety are required." These three traits are closely related. If we could group all of them together, we might call them "tough-mindedness."

Tough-Mindedness

On the track, a driver literally makes hundreds of decisions per lap. Little time is ever available for indecision. Emotions tend to cloud the rational thought process that is required for quick, unambiguous actions. A driver must be decisive, and that demands the ability to control emotions. This is not to suggest that drivers *have* no emotions. They feel love, guilt, pride, anger, and all the rest, in the same way that we all do. The successful driver has learned to separate emotions from rational thought though. For everything, there is a season.

While telling racing stories, one driver explained to me the failure of the steering shaft support in his car while he was in pursuit of another driver. That failure caused his early retirement. Later at the Saturday night beer party, a corner worker who had worked that race told the rest of the story. It seems our erstwhile hero was so frustrated at not being able to make the pass that he was pounding on the steering wheel going down the straights. It was the pounding that broke the shaft support and caused him not to finish. The other driver did not beat him; his own emotions did.

Any driver would be frustrated in such a situation. However, a top driver, when confronted with a situation he cannot control or a competitor who intentionally cuts him off or bumps him, will note the incident in his memory but not make a judgment on it. Rather, he will wait until after the race, when he is likely to explain to the other driver *EXACTLY* how he feels! Dr. Johnsgard continues, "These men have remarkably firm control over their emotions and behavior. It is, in fact, one of their strongest attributes for they possess much more emotional control than the average man, and they must for the race track has hazards enough without allowing impulsive men to drive powerful cars and write their own competition rules."

Some people may feel that they have a great deal of self-control. They should realize, though, that it is much more difficult to control emotions and behavior under conditions of high stress. Racing drivers are generally more composed and relaxed than average motorists when facing high stress levels. Everyone responds differently, of course. When I was driving a number of years ago, I would yawn repeatedly while strapped into the car on the grid. I worked with a driver some years later who had to take medication for nausea before each race. Other drivers become gregarious prior to going onto the track, and some become sullen. All are reactions to stress. It is normal to become nervous before entering the track with 20 or so other drivers who all have the same goal as you. The racing driver's ability to leave these reactions behind, though, is critical to his success.

Dr. Johnsgard considers tough mindedness to be one of the driver's most important qualities. "Sometimes hard, he is often unmoved in situations where most of us would be near hysteria. The world of racing is one of broken friendships, broken machines, denuded bank accounts, and shattered hopes. It demands men who are self-reliant, practical, and responsible. The professional is more tough minded than the novice who is in turn more tough minded than men in general. These are realistic, no nonsense individuals. There is no room on the grid for the tender minded, sentimental, impatient, or demanding male."

Aggression and the Need to Excel

A quality that is extremely important to the success of a race driver is a strong need to be the very best. It is widely accepted among drivers that the race winner was the best on that particular day. He was the best at driving, race strategy, car preparation, and a host of other things. If he is able to win every race, he is the best of all who tried for that period of time. The reason we regard the Grand Prix champions such as Jim Clark, Jackie Stewart, Alain Prost, and Ayrton Senna as above the rest is because they dominated Formula One for their time. They were clearly the best. Drivers who are close friends many times have some fierce on-track battles to prove who is better. "When the green flag drops, the BS stops." Concerning this need to excel, Dr. Johnsgard says that the top professional drivers scored higher in need for achievement than 90% of American college men.

This need to excel can be fulfilled on the track only through a driver's aggression. Especially in club racing, the back of the grid is filled with drivers who have the knowledge, skills, and other psychological characteristics necessary to run up front, but lack the aggression required to charge hard enough to move up. Top drivers tend to be very aggressive and externalize anger.

No group in the thousands of individuals tested by Dr. Ogilvie even comes close to race drivers in scores on aggression and assertiveness. Johnsgard says, "Driver scores on aggression make those of professional football players seem almost docile." Their survey also concluded that it is imperative for these strong aggressive tendencies to be coupled with high levels of the emotional stability previously noted.

How important to the success of a racing driver is his aggression? The Johnsgard study suggests it is paramount. "We've studied a number of drivers who never quite made it big, and often they have all the other personality traits but fall short on aggressiveness." A.J. Foyt in his usual manner is more succinct. "A really good driver can take any car and be competitive by sheer determination alone." No one in his right mind would question the intensity of Foyt's aggression.

Other Factors

Considering the endless frustrations that are an inseparable part of racing, it should come as no surprise that an important trait that drivers share is endurance. Long hours are required in car preparation, races always cost too much, and long trips to and from tracks are frequently disappointing. Races are lost more often than won. The only group of athletes tested by Dr. Olgilvie that scored as high in capacity for endurance was long-distance runners.

Drivers are also very independent. They go their own way and make their own decisions. They tend to be well-informed and skeptical, and they voice their opinions, strongly if needed. To balance their independence, they have a unique ability to accept criticism. This combination of traits keeps the best drivers thinking of themselves as students; they continue to learn throughout their careers regardless of how good they have become at racing.

The quality of independence that drivers share provides a high potential for leadership, but paradoxically drivers show little need to affiliate closely with groups. This may explain why the hierarchy of the sanctioning bodies is composed mainly of administrators. Many experienced drivers would probably not even belong to the club if it were not required. This independence is not characteristic of other athletes. Those involved in team sports tend to be more closely involved with other individuals and groups. Whether the driver's independence is developed because he is alone on the track to do his best in an extremely difficult situation or finds a perfect outlet there, is a question that has not yet been answered.

Ego and the Racing Driver

Each person who drives race cars is different and brings with him a unique set of experiences and a unique value system, which determine how he approaches racing. But these individual personality factors aside, psychologists still are able to group people into a number of categories based on psychological characteristics. At race tracks, drivers are normally only grouped into two categories, which are largely based on how the driver deals with other people in racing and how others should deal with him.

The first group is made up of drivers who are serious about racing and want to do as good a job of it as possible consistent with their goals. They can be found in both club racing and in the pro ranks. They generally are those who have been racing for a considerable length of time or who will be. They are taken seriously by their peers and do not expound a lot of BS. Being part of this group has nothing to do with how competitive a driver is. Serious drivers are concentrated at the front of the grid, but they can be found at the other end, too. Regardless of where they are in the grid, serious drivers tend to group together off the track. If you feel that you should be part of this group, befriend another driver whom you feel is serious and he will (usually) slowly introduce you to his serious friends, providing he thinks you are serious, too. This is also an excellent way of determining who is in the other group.

The second group is made up of "ego racers." These guys race because they want to be big shots, rather than to be successful. (Perhaps that is what success means to them.) I once worked with a driver who boasted to an acquaintance, "I have won 13 races this year!" My reply was, "Mike (or George or José), we've only run seven." He was a Nationally Licensed driver and should have been concerned with National races. Most National drivers use Regional races, if they run them at all, as practice sessions to properly set

The psychological characteristics of a successful driver make him a unique individual and enable him to perform with precision in a hostile environment. Drivers are intelligent, rational, emotionally stable, and accustomed to going their own way and making their own decisions. They are self-sufficient and need little support or approval from others. Johnsgard summarizes his findings this way: "From novice to Grand Prix champion, drivers share a common personality pattern, that rather sharply separates them from other athletes. However, there are differences between drivers of various levels of skill and experience. Considerable selection occurs as men move from novice to established national license holder. That selection seems to favor the emotionally stable, independent, intelligent, tough minded, dominant, venturesome, high-achievement individual."

up their cars for the National events. The National races are the ones that are important to them. This driver, however, wanted to look cool by bragging about his victories—all of them. Even the ones against drivers who were still fulfilling the Regional requirements of their National Licenses.

Other drivers race in classes with little participation so they can brag to their coworkers on Monday morning, "I finished third this weekend." What they fail to disclose, though, is that only two other cars were in their class.

Ego racers are sometimes found in pro racing disguised as wealthy men who have decided they want to race, but have no experience whatsoever. They usually start at the top and work their way down the ladder. The history of racing is full of these guys. They always have the best equipment, but underestimate the importance of the driver's ability to the equation and finish in the last third of the field. They do not usually stay long. This is not to say that all wealthy people who drive race cars are ego racers. Quite to the contrary, some of them hang in there long enough to learn how to be really good drivers and become quite competitive. On the whole, though, rich racers have the reputation, whether it is deserved or not, of being ego racers. This even prompted a novel, published in the '70s, called *Fast Guys, Rich Guys and Idiots.*

Ego racers can be found at the front of the grid and at the rear, even if it is only three deep. They give themselves away by a constant stream of BS which includes heavy doses of brag, not fact. If you are not sure about someone, ask one of the racers you believe to be serious. Many times you will get a response such as, "Joe Blow—pshaw," accompanied by a roll of the eyes.

Physical Condition

> *By doing the right things—getting in peak condition, following smart eating habits, controlling stress, allowing for adequate rest and choosing the best types of training techniques—your chances of success are greatly enhanced.*—
> Mark Martin

Almost everyone who has not experienced driving in competition underestimates the harsh environment and brutal conditions endured by those on the track. Driving a race car is a world of high G loadings throwing the driver sideways in turns, and fore and aft when braking or accelerating, with engine vibrations attacking his legs, back, and hands, and suspension vibrations pounding his entire body as the wheels hit small ripples and rough spots in the pavement. A tremendous amount of physical effort is required for a driver to brace himself against all of these forces. In this environment, it is understandable that many drivers get out of a car after a race, remove helmets, and are seen dripping with sweat.

In view of this, it is surprising that we spend most of our time, money, and attention on improving the performance of the car. The car should be regarded as a tool for the driver to use to win races and, accordingly, he should have the best tool he can use. We tend to neglect the importance of the driver in the system, however. More is to be gained by tuning the driver than tuning the car. If we can enable him to be more alert, have more stamina, increase agility and coordination, and decrease reaction time, he will be better able to cope with the demands of the track and can do so for longer periods and with greater efficiency. When someone is fit and alert, he is a faster, safer driver. When a driver is fatigued, his performance suffers and he will be slower as a result.

Being in shape also means that the driver is carrying less unnecessary personal weight that the car must accelerate and decelerate, improving performance. It is difficult and expensive to get weight off the car, but comparatively easy to remove it from the driver. It is rare to see a driver on the podium who is overweight.

The major benefits of fitness to a racing driver are improved coordination, reaction time, alertness, endurance, muscular flexibility, cardiovascular efficiency, respiration, and heat tolerance, as well as lessened injury severity in the event of a crash, and decreased recovery time. An additional benefit of fitness is better sleep. Sound sleep is not only a result of an intense workout, but is also a lasting benefit of a long-term exercise program. Interrupted sleep or insomnia is something that cannot be tolerated if maximum performance is expected.

Racing produces a multitude of stressful situations that can build on those from our normal lives, ranging from meeting deadlines to dealing with officials. Stress is something we all experience and, if it is chronic, it can become debilitating. Studies have shown that those who are not physically fit have the least tolerance to stress. Even if stress is not excessive, it can be detrimental to on-track performance. Stress has a negative impact on energy levels and the ability to recover from injury and illness. In addition to being better able to deal with stress if you are fit, exercise provides an avenue through which frustrations can be vented.

All of these benefits of fitness to the driver do not come easily, and not all physical activity is a step toward accomplishing a driver's goals. Becoming fit is a matter of regular, intensive workouts, and not all exercise programs are created equal. Different forms of training will produce different results. The ones in which we are interested for racing performance include improvement in the cardiovascular and respiratory systems and in the muscular strength and agility of specific muscle groups.

Aerobics

Aerobic activity increases cardiovascular efficiency and breathing performance. It helps to promote overall fitness. If you are not currently in good overall condition, aerobic exercise is the place to start. During exercise, the heart rate increases, and the contraction of the heart is greater, pumping an increased quantity of blood. At the same time, the arteries and vessels expand, helping to increase the delivery of oxygen and nutrient-rich blood to the tissues. The need for additional oxygen during exercise makes the lungs expand more and do so more frequently. This action is controlled by the diaphragm. All of these muscles are strengthened by aerobic exercise so that they will be more easily able to cope with extra demand when necessary. When demand is above normal but below exercise levels, as when driving, the heart and lungs will not have to work as hard to supply oxygen and nutrients to the body's tissues.

A sustained, elevated heart rate is generally considered to be the best aerobic exercise to increase cardio-respiratory fitness. Exercises that will accomplish this include running, swimming, jogging, cycling, and cross-country skiing. Many exercise machines are available to simulate these activities so that any exercise may be continued year round. Most health clubs also have aerobics classes. If all else fails, there are always exercise videos.

Mark Martin, one of the top NASCAR drivers, is widely known as a fitness expert. In collaboration with John Comereski, a highly regarded exercise physiologist, he has recently published what is destined to become the physical training "bible" for racers, *Strength Training for Performance Driving.* In regard to the importance of aerobic training, Martin says this: "Cardiovascular or aerobic conditioning pertains to the ability to engage in active physical activity for a sustained period of time, without becoming fatigued. For the most part, it involves your heart, arterial network (arteries and veins), lungs, and blood. It is the most important concern for health and fitness."

Although Martin emphasizes the importance of cardio-respiratory fitness, aerobic exercise is not the only way to attain it. He continues: "[A] growing number of research studies has shown interval strength training to be a productive means of aerobic conditioning. Interval training raises your heart rate during intense work periods, then lowers it during passive periods. So your heart rate rises and then declines, and so on. This can best be accomplished through circuit training."

Strength Training

> *You can have the fastest car on the NASCAR or Indy circuit, but if you are not capable of driving it intensely, without becoming unduly fatigued, you might as well stay home.*—Mark Martin

When a training program is composed entirely of aerobic activity one concern is the utilization of nutrients in muscle tissue. Metabolism of nutrients within a cell requires oxygen and metabolizing enzymes, which are produced and used in local areas, as they are needed. These enzymes are produced in larger quantities during times of intense physical activity. Aerobic activity, of course, is generated by muscle contraction, and the cells of the specific muscles involved have the enzymes available that are required for metabolism. However, the muscles not involved in the activity have only nutrients and oxygen, and merely store the nutrients as fat. It is important that all of the muscles specific to driving be exercised for those muscles to be properly toned and strengthened. Aerobic activity is very good for cardio respiratory fitness, but the specific muscle groups involved in driving a race car must be exercised, too.

A race driver should exercise a total of 14 specific muscle groups. This number is much higher than that recommended for someone only interested in general conditioning. Before you begin, look for an exercise physiologist or trainer to design a conditioning program suited to your needs. Make sure he knows the purpose of your training and understands the rigors of racing. Buying him a copy of Martin and Comereski's book would be a good start. When beginning a workout program, it is neccessary to get a complete physical—not the physical required by the sanctioning body for your racing license, but a real comprehensive physical by your own doctor. Take your doctor a copy of your workout regime so that he knows exactly what you will be doing. Some communication between your trainer and your doctor may be necessary.

MUSCLE OR MUSCLE GROUPS TO BE STRENGTHENED	
Wrist Flexors	Posterior Deltoid
Biceps	Latissimus Dorsi
Triceps	Spinal Erectors
Quadriceps	Abdominals
Pectorals	Hamstrings
Anterior Deltoid	Gastrocnemius
Lateral Deltoid	Trapezius

Stretching and Warming Up

Muscles should be worked lightly before beginning a workout. A warm-up increases blood flow to the tissues and increases body temperature. This rise in temperature is a result of increased metabolic action which, again, means oxygen and nutrients are more easily assimilated into cells. This is important when a more strenuous workout is begun. Additionally, as temperature goes up, the muscles become more limber and their contractions quicker and more forceful. Increased body temperature also aids in the transmission of nerve impulses into and through the muscles allowing quicker reactions.

Prior to a workout session, you should warm up using the same exercises you will be using during your work out, but with very light weights and a higher number of repetitions. This will warm up the specific muscle groups to be exercised. The warm up should be continued for at least five minutes or until intramuscular temperatures rise slightly. Aerobic exercise is an acceptable alternative to circuit warm ups, but not as good for specific muscle groups.

Along with warming up, it is also important to stretch the muscles, tendons, and the connective tissue in the joints. Stretching increases flexibility while decreasing the possibility of injury during the workout. It will also decrease soreness after a workout. Minute tears in the muscle fabric caused by exercise cause soreness. Stretching helps to limber up the muscles and prevent this tearing.

Stretching the specific muscle without load and holding the position for eight to ten seconds is the way it should be done. Release the muscle and relax, and then repeat as many times as necessary. Another method which can be used concurrently with the warm-up is to use the full range of motion in the warm-up exercises so that at the end of each stroke, as one muscle is contracted, others are being extended and stretched. Be careful not to stretch a muscle that is under load. This can injure the muscle, doing more harm than good.

Developing a Training Program

> *If the race driver and crew spend their exercise time jogging, it's our belief they may as well take a snooze.*—Mark Martin and John Comereski

Three distinctly different types of training programs should be considered for conditioning depending on your goals and your present condition. These are: general conditioning, strength conditioning, and speed/power conditioning. If you are not currently involved in a conditioning program, they should be taken in this order. Begin a general fitness program and continue it until you attain a good overall fitness level. Then move on to strength training. Speed and power training should begin after some strength training, when you are ready for it.

A general fitness program will consist of lower levels of exercise and possibly fewer muscle groups than what we are really interested in for peak racing performance. It is always better to start slow and build up your exercise program gradually by adding muscle groups as you are ready. A general rule is that as weight and exercise speed go up, fewer repetitions are required. A general program should include 8 to 12 repetitions of each exercise, with strength training falling to six to eight repetitions, and speed/power training dropping to two to six reps in some cases. For strength and speed/power conditioning, the amount of weight should be selected to generate muscular failure, i.e., to the point that you cannot lift the weight in any further repetitions. Begin with the minimum number of repetitions for each type of training and add reps as you are able. When you reach the maximum number of reps, increase the weight five to ten percent. The same principle applies to general conditioning, except that the exercise should not be taken to muscular failure. When you reach a plateau and improvement seems difficult, it is a signal that your program should be altered. Regardless of the training type, be sure to take each repetition to the maximum limits of its stroke. This will ensure that the muscle and tendon are strengthened completely to their limits of contraction.

Muscles need time to recover between workouts. Different types of muscles heal at different rates, but almost all recovery times fall within 48 to 96 hours. For this reason, it is wise to time your workouts two to four days apart depending on the muscle groups exercised and the intensity of the workout. The quadriceps, hamstrings, and muscles of the chest and back are the slowest to recover. The neck muscles, those of the calves and forearms, and the abdominals are the quickest.

These are only guidelines, not hard and fast rules. Women, for example, will usually use more repetitions and less weight. Enlist the help of a qualified trainer and follow his advice. He can help to design a program tailored specifically to your needs. During the course of your fitness program, the trainer can help when you come up against difficulties and obstacles. Remember that no exercise program will benefit your overall health or increase your racing performance unless you use it regularly.

Motivation

All drivers may have the same motivation to work out, i.e., to improve overall driving performance, but each person has slightly different goals. Some may want to look better, while others want to feel better or be stronger. Goal setting is extremely important to continuing an ongoing fitness program. Long-term goals such as these are beneficial, but too vague to motivate you to work out on a regular basis. Sub-goals are needed so that you have something more tangible to work with. Increasing the amount of weight to be lifted or number of reps could be set as sub-goals. And when you reach a goal you can give yourself a pleasurable reward, such as a special evening with a special lady or that new widget you want for the race car. Motivation is easy when the sub-goals are realistic and the rewards are ample.

Sample Workouts

Following are some outlines of *sample* workouts. Many variations are possible, and your specific conditioning needs should be taken into consideration before deciding on a particular regime. These programs are designed for men. For women or for people with special needs, please consult your trainer. The strength and power/speed workouts are both two-day programs designed to work different muscle groups on successive days to give them time to heal. These are pretty intense workouts at each level. They are intended to show what is possible and what you should strive for in a workout program. Have your personal trainer design similar programs for you and follow them. Step up from a general conditioning workout to a strength-training workout when you are ready, and then to a power/speed workout. When you are ready to graduate from one of *those,* your competitors will not want to meet you in a back alley or on the race track.

GENERAL CONDITIONING WORKOUT

Exercise	Sets	Reps
Reverse Crunch	2	16–20
Twisting Crunch	2	16–20
Roman Chair Crunch	2	16–20
Incline Crunch	2	16–20
Knee Up	2	16–20
Leg Curl	2	8–12
Leg Extension	2	8–12
Fly	2	8–12

GENERAL CONDITIONING WORKOUT *(cont.)*

Exercise	Sets	Reps
Upright Row	3	8–12
Pushdown	3	8–12
Seated Row	2	8–12
Pulldown	3	8–12
Dumbbell Curl	3	8–12
Standing Calf Raise	1	16–20

STRENGTH TRAINING WORKOUT

Day One

Exercise	Sets	Reps
Reverse Crunch	2	16–20
Knee Up	2	16–20
Twisting Crunch	2	16–20
Roman Chair Crunch	2	16–20
Incline Crunch	2	16–20
Leg Press or Squat	4	6–8
Leg Extension	3	6–8
Dumbbell Bench Press	4	6–8
Fly	3	6–8
Overhead Press	3	6–8
Lateral Raise	3	6–8
Bent Over Lateral Raise	3	6–8
Shoulder Shrugs	3	6–8
Seated French Curl	4	6–8
Triceps Dips	3	6–8

Day Two

Exercise	Sets	Reps
Reverse Crunch	2	15–20
Knee Up	2	15–20
Twisting Crunch	2	15–20
Roman Chair Crunch	2	15–20
Incline Crunch	2	15–20
Leg Curl	3	6–8
Stiff Leg Dead Lift	4	6–8
Chin Ups	3	6–8
Pull Ups	3	6–8
Seated Row	3	6–8
Good Mornings	3	6–8
Dumbbell Curl	3	6–8
Preacher Curl	3	6–8
Reverse Curl	3	6–8
Wrist Curl	3	6–8
Standing Calf Raise	3	16–20

INTEGRATED STRENGTH AND POWER/SPEED WORKOUT

Day One

Exercise	Sets	Reps
Leg Press or Squat	3	2–6
Leg Extension	2	6–8
Leg Curl	2	6–8
Stiff Leg Dead Lift	4	2–6
Reverse Grip Pulldown	2	2–6
Pulldown	2	6–8
Seated Row	2	6–8
Good Mornings	4	2–6
Standing Calf Raise	3	16–20
Reverse Crunch	2	16–20
Knee Ups	2	16–20
Twisting Crunch	2	16–20
Roman Chair Crunch	2	16–20
Incline Crunch	2	16–20

Day Two

Exercise	Sets	Reps
Bench Press	4	2–6
Dumbbell Bench Press	2	6–8
Overhead Press	3	2–6
Upright Row	3	6–8
Bent Over Lateral Raise	2	6–8
Shoulder Shrugs	3	6–8
Pushdowns	2	2–6
Seated French Curl	3	6–8
Dumbbell Curl	3	2–6
Hammer Curl	3	6–8
Wrist Curl	3	6–8
Reverse Crunch	2	16–20
Knee Ups	2	16–20
Twisting Crunch	2	16–20
Roman Chair Crunch	2	16–20
Incline Crunch	2	16–20

Nutrition

It may seem much too basic, but to be in good physical condition you should not only eat, but also eat properly. Conditioning or strength training will be of little value if your diet does not provide adequate nutrition. Oxidation occurs in cells due to their natural function, and the rebuilding of those cells requires nutrients. As you stress your body through exercise or the activity of driving a race car, the demand for proper nutrition increases proportionally. In addition to the overall health benefits, drivers need to practice good nutrition for these six basic reasons.

1. **Fat Loss**
   We seem to spend countless hours and untold dollars trying to rid our cars of unwanted weight. It is usually much easier and cheaper to remove it from the driver.

2. **Increasing Strength**
   Protein plays an important part in strength improvement. It is protein that increases muscle size and, therefore, strength. No matter how much you exercise, if your diet contains inadequate protein levels, muscle tone will not improve and strength will not increase.

3. **Energy Increase**
   When muscles are used intensely, sugar in the form of glycogen in the muscle tissue is used as fuel. It is replenished by glucose in the blood which, of course, comes from the food you eat. Although many foods contain sugars, they have different glycemic ratings. The lower the glycemic rating, the more long-duration energy supplies the food contains.

4. **Increasing Mental Performance**
   Sufficient intake of carbohydrates is essential to mental processes, just as it is to proper muscle function. Without ample carbohydrates, the transmission of neurological impulses slows, causing impaired attention and ability to concentrate, as well as a feeling of fatigue and weakness in the same way as would lack of sleep.

5. **Recovery from Exercise**
   Regardless of whether the exercise in question is strength training or racing, the muscular activity results in cellular oxidation and minute tearing of muscular tissue. The repair that the body naturally performs on these tissues uses protein. Additionally, the depleted reserves of glucose and glycogen must be replenished, which requires carbohydrates. Improper nutritional habits lengthens the time required for the body to perform these functions.

6. **Recovery from Injury**
   Although none of us likes to think about the possibility of a racing injury, injuries can and do occur, and do so without warning. Proper nutrition speeds recovery time by providing carbohydrates and proteins. Carbohydrates provide the necessary energy reserves and proteins provide amino acids, which are the building blocks of human protein. Protein

also enhances the immune system to protect against infection during recovery. Vitamins and some minerals are also responsible for reducing the effects of fatigue that accompany the recovery process.

It should now be clear just how important nutrition is to a racing driver. Five major nutrients are necessary for proper nutrition: carbohydrates, proteins, fats, vitamins, and minerals.

Carbohydrates

Carbohydrates are the body's fuel. Just as a racing engine requires fuel of a specific quality, some carbohydrates are of better quality or are more appropriate for the body than others. The three

Age and the Racing Driver

What is the optimum age for a driver? The answer to this simple question is not a simple one, as always seems to be the case in racing. In the European manner of thought, a driver is washed up by the time he is 35. Although there are some exceptions, few drivers in Formula One (F/1) continue to race past that age. The F/1 teams seem to look for young lions who have graduated from Formula 3000 at the ripe old ages of 21 to 25.

In North America, we tend to have a more realistic approach, whereby the value of a driver is based more on what he can do on the track and less on the color of his hair. We do still recruit some young drivers into top level motorsports. Michael Andretti and Al Unser, Jr. both drove Indy cars at tender ages. More recently, Paul Tracy has done the same. Other young drivers who have recently entered the top level include Jeff and Robby Gordon, Brian Till, Brian Herta, and Jimmy Vasser, each of whom was in his twenties when he first got a shot at the big time. An interesting observation is that during the first half of the '95 Indy Car season, Jacques Villeneuve had won two races at the age of 24, Robby Gordon two at 26, and Tracy two at 26, also. The other two races were won by Unser, a veteran at 33, and Fittipaldi at 48.

Some drivers on this continent continue to drive well into their forties and even fifties. Mario Andretti's recent retirement from Indy Cars came at the age of 52, and he is still doing selected sports car races. Emmerson Fittipaldi retired also at the age of 48. A.J. Foyt drove until the age of 54. Paul Newman won the 24 Hours of Daytona (with help from some other very qualified drivers) at the age of 70!

The proper age for a driver to move into the top levels of motorsport seems to be as soon as he is able to handle it. This includes having the necessary car control skills and proper judgement. Some young drivers have one but not the other, and do not usually last long unless they quickly develop the missing quality. At the other end, a driver should continue to race as long as he is categories of carbohydrates are monosaccharides, disaccharides, and polysaccharides. All of these are sugars, and sugar is the most concentrated form of energy that is easily assimilated into the body. The mono- and disaccharides are simple sugars consisting of glucose, sucrose, fructose, and lactose. These sugars are found in fruits, honey, and milk. Sucrose is normal table sugar.

Polysaccharides, sometimes called complex carbohydrates, are the body's best source of fuel. They are stored in food as starches and must be converted to sugars (cellulose, glucose, pectin, and dextrin) by the digestive system. The blood then carries these sugars, transferring them to the tissues, where a certain amount is stored as glycogen. Any excess will be converted to fat. The glycemic index of a carbohydrate is a measure of how quickly it is converted into these blood sugars and how long they are available in the blood to be converted to glycogen by cellular action. The lower a food's glycemic index, the longer lasting is the fuel it provides to the body.

competitive and enjoys it. Some drivers, however, enjoy it past their competitive days. Of these drivers, Foyt says, "Racing is a strange business. A lot of guys who should make it don't, some who shouldn't do, and a whole lot of others stick around too long. They may have had it at one time or another, but for some reason—age, loss of nerve, or worse yet, loss of determination—they slow down. And rather than quit, they just keep going, dragging it all down the other side."

There is no optimum age for a race driver. The sport does favor the young, but the benefits of youthful enthusiasm, eyesight, and reflex action must be tempered by experience. A popular bumper sticker in paddocks reads, "Old age and treachery will overcome youth and exuberance."

A related question is, "How much experience does it take for a driver to become proficient at the highest levels of the sport?" As with the previous question, it depends. The answer to this question may be at least a little easier to pinpoint, however. Many of the current era's top drivers, including Ayrton Senna, Al Unser, Jr., and Paul Tracy, began by racing Karts at the age of 10 or 12. By the time they were of voting age, they had over a decade of racing experience and were just beginning to get a taste of the big time. Jimmy Vasser won the SCCA Formula Ford National Championship in 1986, and in 1995 took his first Indy Car victory. Michael Andretti had raced for a number of years before winning his first Indy Car race. And who can forget Al Unser, Jr. his first time at Indy, while a lap down at the finish, blocking Emmerson Fittipaldi so that his dad could have a better shot at winning. A few years later, he wasn't a lap down at the finish. The truth is, no one can predict how much experience any driver will require before being competitive at the highest levels. One generalization can be made, though: that time is measured in years. I once heard about a would-be driver who, having never raced anything except at Malibu Grand Prix, while looking at his first car to buy, is reported to have said, "I wouldn't rule out Formula One in a couple of years." Well, friends, except for Gilles Villeneuve, it just doesn't happen that way. If you plan to run NASCAR, Indy Cars, or Formula One, plan to practice for a while.

A large intake of any sugar is not wise, but this is especially true of the simple sugars. The body combats a large and sudden rise in blood sugar level by increasing insulin production. The insulin breaks down excess sugar in the blood leaving the blood sugar level lower than it was before the sugar intake. So, although it was common practice years ago, you should not eat a candy bar shortly before each race. Instead, the majority of each meal, with the possible exception of the last meal of the day, should be made up largely of complex carbohydrates. These complex carbohydrates provide long-lasting energy stores and are available to the tissues when needed. By not consuming large amounts of complex carbohydrates shortly before bedtime, you are decreasing the amount of sugars that will be turned into fat during inactive periods.

Protein

It is commonly known that most of the body's weight is water. The remaining dry matter of the body is almost half protein. There are a number of different types of proteins in the body, but all of them are composed of amino acids. About half of these amino acids are manufactured by the body's natural processes, whereas the others must be consumed as dietary proteins. When a deficiency exists in these amino acids, the growth and regeneration of the cellular structure is inhibited. However, some animal proteins do not contain the amino acids that the body requires. Proteins are rated by their quality or the number of essential amino acids they include. Low-quality proteins include all fruits and most vegetables. Meat, fish, and poultry are much higher-quality proteins, and milk and eggs are considered complete proteins. It is little wonder that nature provides complete proteins for the growth of the young.

When athletic activity increases, so does the body's need for protein. It is essential that sufficient quantities of the appropriate amino acids be present when required for growth. This is usually during and immediately following exercise. Actually, it is best if some quantity of protein is included with most meals, but it is especially important that protein be included in the last meal eaten before a workout or race.

Although it is imperative that sufficient quantities of protein be consumed, it is possible to include too much protein in your diet. Extremely high protein intake can strain the liver and kidneys. Excessive protein can be converted to fat by the metabolizing enzymes and stored by the cells. The all-protein diets that are currently in vogue omit some nutrients that are vital to good health. Regardless of what you might have heard, these diets should be followed for no more than four days at a time and only during periods of inactivity.

Fats

While carbohydrates are the body's primary source of energy, and protein is used for cellular regeneration, fats are also necessary for proper function. Fats aid in many bodily functions including regulation of cholesterol, and are also carriers for certain fat-soluble vitamins such as A, D, E, and K. Fats are a secondary source of energy, which are called upon when adequate energy from carbohydrates is not present. Fat contains a very high concentration of energy, but is not as easily used by the body as carbohydrates. Most fats contain about nine calories per gram, which is about twice that of both carbohydrates and proteins. This is why foods high in fats are also high in calories. Fat is an essential component of the body and an essential part of the diet.

Fats come in two basic types: saturated and unsaturated. Saturated fats are usually animal fats in meat and poultry as well as dairy products. Most vegetable oils do not contain saturated fats, but the two exceptions are palm oil and coconut oil. Unsaturated fats contain much lower levels of cholesterol and are found in other vegetable oils. The unsaturated fats come in two varieties, too: monounsaturated and polyunsaturated. Of these two, the polyunsaturated oils contain less undesirable cholesterol. However, vegetable oils are sometimes hydrogenated, a process that increases the shelf life, but also basically saturates unsaturated fats. Hydrogenated vegetable oils are usually found in cooking oils, particularly those that are solid at room temperature such as margarine and shortening.

Fat is a necessary part of the diet, but it is so easy to consume too much of it, that a concerted effort should be made to limit fat intake. This is not only because of the need to control body fat, but also to limit the cholesterol circulating in the blood. Cholesterol also comes in two types: LDL and HDL. Both are manufactured by the body and are required for cellular regeneration. LDL is generally known as "bad" cholesterol. When too much saturated fat is included in the diet, excess LDL cannot be processed and circulates with the blood where it eventually settles out, clinging to arterial walls and reducing the size of the artery, thus restricting blood flow. HDL is "good" cholesterol and attaches to the LDL, removing it to the liver for reprocessing or removal. Saturated fats are the primary source of LDL, but smaller quantities are present in unsaturated fats. Unsaturated fats generally contain HDL as well. For this reason, the fat content in the diet should be primarily unsaturated fat. (Other foods are said to reduce LDL levels too, such as red wine and garlic.) The recommended dietary intake of unsaturated fat is 10 to 15 grams per day for a body weight of under 200 pounds and 20 to 25 grams per day for a body weight over 200 pounds. In practice, with western diets, this is *very difficult*! Reading labels on packaged food reveals that many contain over half of these amounts each. A typical breakfast pastry, for example, may contain 25 grams of fat. Lowering fat intake is a matter of being conscious of fat content and the type of fat the food contains, and having the willpower to eat healthy foods that are low in fat. A recently introduced product called Chitosan helps to reduce the fat content of the body. Chitosan attaches to fat particles in food, but is indigestible so it, along with the fat particles, passes right through the system. Chitosan has been shown to lower cholesterol levels as well. Just as we are weight-conscious when building or modifying a race car, we can be fat-conscious when eating. An apple with that morning cup of coffee will make you faster; a donut will not.

Vitamins

Biologists have discovered many different vitamins, and each plays an important role in the body's function. Most have to do with the metabolism of sugars at the cellular level. However, other vitamins assist in the growth of red blood cells, management of cholesterol, blood clotting, immune system function, neuron transmission, and many other functions. The major vitamins responsible for these functions are A, B_1 (thiamin), B_2 (riboflavin), B_3 (niacin), B_5 (pantothenic acid), B_6 (pyridoxine), B_{12} (cobalamin), C (ascorbic acid), D (calciferol), E (tocopherol), K (menadione), and folic acid. A well-balanced diet will usually include some quantities of each of these vitamins. They are found in fruits, vegetables, grains, dairy products, meat, poultry, and fish, in other words, in all the food groups. However, there is no guarantee that a diet including ample carbohydrates and protein also includes sufficient quantities of vitamins. Some nutritionists now believe that because of farming soil depletion, our fruits, vegetables, and meats no longer contain sufficient vitamins to ensure proper health with normal sized meals. Advertisers may lead us to believe that taking one miracle pill per day will do the trick, but the reality is that it takes much larger doses of specific vitamins to obtain optimum health. Vitamin programs have been developed by nutritionists, anti-aging doctors, and the compounding pharmacies that supply them, to enhance endurance, mental function, weight loss, and other functions. These are available through several companies, including one called Healthy Living, Inc. (See Appendix C for more information.) Except for the fat-soluble vitamins (A, D, E and K), overdosing on vitamins has generally little or no detrimental effect. I take 1000 mg each of vitamins E and C daily, primarily because they are very good anti-oxidants. Taking megadoses of extra vitamins arbitrarily, though, is at best a waste of money and at worst can affect health adversely.

Minerals

Like vitamins, minerals are also important to body function in a myriad of ways. They assist in bone growth, red blood cell formation, kidney function, muscle function, and several other functions. The major minerals associated with these duties are potassium, magnesium, calcium, sodium, iron, and phosphorous. These are found in numerous foods and in tap water, as well as mineral water. However, iron and potassium are two minerals that seem to be in short supply in American diets. A mineral supplement containing these two may be a good idea.

Another good source of most minerals is sports drinks, Gatorade, 10K, and the like. Unlike vitamins, however, massive quantities of minerals can have severe side effects. It is well known that too much salt (*sodium* chloride) can elevate blood pressure. Excessive quantities of other minerals have similar severe effects, such as irregular heart action, nausea, kidney action impairment, and buildup of toxic substances in the liver. Therefore, it is wise to closely control the intake of extra minerals. Sports drinks are excellent when perspiration rids the body of some of these essential minerals, but their use should be limited to those occasions. Under normal circumstances ordinary water is the best drink.

The best method of properly controlling mineral intake is through an analysis of your hair. Blood and urine tests are very good indicators of what substances are in the body, or have been in the body during the previous 24 hours. However, an analysis of a person's hair will show which mineral, in what concentrations, are part of the body over a longer term, usually a period of several months. Some minerals in high concentrations can be toxic. High levels of aluminum, for example, are always found in Alzheimer's patients. Where do you get high levels of aluminum? From drinks contained in aluminum cans, from aluminum cookware, and from antiperspirants. (And what about all of those race car fabricators who work with aluminum all the time?) Other minerals can cause problems as well, both when they are present in toxic levels and when there are deficiencies. Hair tests can show these levels and indicate what dietary changes should be made to achieve the proper balance of minerals.

Water

Although not a nutrient per se, water is another substance that the body requires for proper nutrition. Mammals are made up of almost 90% water, and a slight loss of body fluid through perspiration or excretion can have a significant effect on all the ordinary body functions in which the nutrients are used. Water is the solvent in which the other nutrients are dissolved. It is imperative that sufficient water be included with the diet each day. This can be in the form of coffee, tea, fruit juice, milk, or other liquid foods, which are made up primarily of water, but you should remember that all of these include various sugars, amino acids, and other compounds that will affect your body in their own ways. It is okay to drink fruit juice or other drinks when you feel you need that particular food, but it is best to drink water when you feel the need for additional fluid.

It seems that proper hydration is especially important *prior* to intense exercise. One of our worst habits as humans is to not partake of sufficient water until we become thirsty. We think we are camels! In her book, *Healthy Healing*, Linda G. Rector-Page, N.D., Ph.D., explains it this way: "Water is second only to oxygen in importance to health. Thirst is not a reliable signal that your body needs water. Thirst is an evolutionary development designed to indicate *severe dehydration* [emphasis is the Doctor's]. You can easily lose a quart or more of water during activity before thirst is even recognized. Plain or carbonated water is best. Second best are unsweetened fruit juices diluted with water or seltzer and vegetable juices. Alcohol and caffeine drinks are counter-productive because of their diuretic activity. Drinks loaded with dissolved sugar or milk increase the need for water."

To drive a race car well, you must be sensitive to your car and feel all of the subtle nuances of its behavior. To be in proper *condition* to drive a race car well, you should be as sensitive to the needs of your body. You should be able to discern when you need fluids, when your workouts are too intense or when they are too close together, and when you need to eat carbohydrates or protein. We have each lived in our bodies for many years, but most of us do not pay enough attention to things our bodies tell us. Listen to your body and it will give you valuable information you can use to improve your performance on the track.

Hormones

Although they are neither vitamins nor minerals, hormones are also vital to the proper function of the body. The medical profession seemed to consider them mysteries for decades, but numerous studies during the '80s and '90s have revealed what hormones do and how they work. The major hormones are melatonin, DHEA, testosterone, estrogen, progesterone, and growth hormone. Although all of these have important functions in the operation of the body, studies completed in July of 1998 have concluded that growth hormone or HGH is instrumental in the formation of the others. HGH is produced by the pituitary, a small gland located in the brain. The production of HGH by the pituitary peaks around the age of 18 years, then starts a slow decline to the age of 30, when it drops sharply. By the age of eighty, HGH production by the pituitary is around six percent of the peak production at the age of 18.

HGH not only promotes the growth of bones and muscle a child requires to form the body of an adult, but it is also necessary to prevent the atrophy of every part of the body which we call aging. Increasing the HGH levels in the body enhances the function of every system, including increasing muscle mass, decreasing fat, increasing immune system function, enhancing cardiovascular system efficiency, enhancing eyesight, increasing respiratory system function, increasing sexual ability, and increasing the size and functional ability of all the organs of the body including the brain, kidneys, and liver. Interestingly, cholesterol clogging of the arteries, which causes a large number of heart attacks each year, has now been linked to poor liver function. One of the jobs of the liver is to eliminate excess LDL cholesterol from the blood. Exhaustive medical research on HGH is now showing this hormone to be the fountain of youth for which Ponce de León was searching. The general health benefits of HGH are obvious, but its specific benefits in extending driving careers may lead to an increase in the average age of the field during the first two decades of the twenty-first century.

Dr. Ronald Klatz, a medical doctor who is one of the front runners in the race to find the causes of the aging process, says this in his book, *Grow Young with HGH*: "Human Growth Hormone is the first substance that has been clinically shown to actually reverse the effects of aging. It has the potential to change the way we live and grow young ... The reality for most of us who are alive today is an average life span in excess of 100 years with unexpectedly good health. Fifty years from now when the seventy-six million baby boomers start reaching the century mark, we will look back on the medical science of today as if it were the dark ages. A healthy and athletic 105 year old will be like a fit and active 65 year old of today, while the 65 year old of the future will be someone who is now 35 years old."

Although experience and judgement may allow older race drivers to make up for reduced reactions, coordination, stamina, eyesight, etc., keeping them on equal footing with the younger drivers, it would seem wise to keep the body young and be able to use the experience and judgement along with youthful reactions, coordination, stamina, eyesight, etc. Would you want to race against someone like A.J. Foyt, who at 65 had fifty years of racing experience and the body of a 35-year-old? No? Well, then, would you like to *be* a driver with that much experience and a young body?

When and How to Eat

Now that we know the importance of the nutrients that make up the foods we eat, we need to get down to some practical rules for when and how to eat. As mentioned before, the body's nutritional needs vary with its activity level. Many trainers recommend eating several smaller meals five to six times a day. This means that the planned activity and its intensity should be considered before determining what to eat and how much to eat. If a very active period is anticipated, complex carbohydrates should make up most of the meal. When less strenuous activity is anticipated, a meal made up of complex carbohydrates and protein is the proper choice. If another meal is planned in three to five hours, it should be obvious that it should not be a large one.

We can learn many things from observing nature. The body's need for protein, for example, means that humans are carnivorous creatures. We still have the remnants of incisors to prove it, just as other carnivorous primates do. Many of these primates also eat plants, as we do, and their "meals" consist of numerous small feeding periods throughout the day. Although we may feel as though we have elevated ourselves above the animal kingdom, we still have similar genetic makeup and similar nutritional needs. Other primates do not have the social schedules we do, so nature is the driving force in their lives. They eat when and how much is necessary. Perhaps we could learn a lot from a monkey.

A typical day for a fitness-oriented racer may go something like this: Breakfast, consisting of a fruit plate and fruit juice or coffee, is followed mid-morning by a snack of low- or non-fat yogurt. Lunch may consist of a taco salad made up of greens, lentils, tomatoes, olives, a little chicken, and a slight amount of low-fat dressing. An early dinner of good portions of pasta with eggplant and a small amount of ricotta cheese is followed by an early evening workout. A meal such as grilled salmon should be eaten an hour after the workout and a couple of hours before bedtime, to once again build up protein stores. This schedule should serve only as a guideline. Your daily schedule, including when you work out and the type of work you do during the day, will affect when and how you should eat. Develop your own good eating habits and stick to them. In spite of those high-protein diet gurus who insist that starchy carbohydrates make you fat, Alex Zanardi maintains his weight on a diet of pasta almost every day, along with lots of vegetables, fruit, bread, olive oil, and little meat—a classic Mediterranean diet. Alex explains, "The weight of a driver is a big issue. Gaining even a few pounds can make you lose a tenth of a second per lap, an advantage you don't want to give away to the competition." The following guidelines may help.

1. **Eat Breakfast.** It's important to raise your blood sugar level in the morning.

2. **Eat lightly** before relatively inactive periods.

3. **Eat low-fat foods.**

4. **Eat fresh foods** as much as possible. Avoid processed foods.

5. **Eat larger meals composed of complex carbohydrates** from the lower end of the glycemic index scale prior to active periods.

6. **Eat meals made up primarily of protein as the last meal of the day,** at least two hours before bedtime.

On a race weekend, the same rules should apply, but your schedule is likely to be rearranged. The night before a day of driving activity, a protein meal should be taken. The next day, breakfast should consist of whole-grain pancakes or cereal accompanied by fruit or fruit juice. During the day, eat additional small meals of fresh fruit, which can be kept in an ice chest at the track. Drink plenty of water and some sports drinks. Avoid soft drinks. Eat another protein meal that night when you go out with the crew.

Coordination and the Senses

The obvious physical traits such as eyesight and coordination are important to be sure, but a deficiency or limitation in these areas is no reason that a driver cannot be competitive at any level. Bobby Rahal, Brian Herta, and Paul Tracy all wear glasses when they drive. Mel Kinyon lost several fingers in a fire many years ago and still drives Midgets using a special glove fitted with a socket attached to a stud on the steering wheel. To make the job easier, though, eyesight and coordination should be optimized. Your eyesight should be correctable with glasses or contacts. If you are able to pass an eyesight test for a state driver's license, you can most likely see well enough to race.

The focusing of the eye to differing focal lengths is a different matter, though. In order to focus on an object, the eye changes the curvature of its lens, a process called accommodation. A young person can focus to a very short distance and can do so very quickly. However, as we age both the minimum focal distance and the rate at which the focus can be changed increase. The rate of focus can be as much as 1/2 second and a racing car can travel quite a distance in that time. In addition, both factors are important with regard to the placing of instruments in a race car. They should be mounted as far away from the driver as possible with large, clear markings to make focusing on them quick and easy.

The clarity of an image that is in focus on the retina of the eye is measured by visual acuity. This is the factor that is measured by the optometrist's eye chart. Normal visual acuity is considered to be one minute (one sixtieth of a degree) of arc at the eye, corresponding to about one and a half nerve cells at the retina, which would seem to be a theoretical limit. A few people seem to have better visual acuity than this, a phenomenon that is not yet completely understood.

A trait that is important to a racing driver is the rapidity with which he can make use of his visual acuity. If an object that is normally just barely perceptible is visible for a short time, one second, only about 50% of the populace will recognize it within this time. The rest will not be able to use the limits of their acuity. Some ophthalmologists feel that this is a perceptive function within the brain.

In the same way the overall condition of your body can be improved through exercise, specific parts such as the eyes can be made to function better, too. Exercising the tiny muscles that affect the rate and distance of focus will make them stronger and more efficient. Some doctors feel that these activities can also benefit the perceptive abilities of the brain. One program that seems to be effective is a book called *How to Improve Your Vision without Glasses or Contacts.* (See Appendix B for more information.)

Some drivers who use corrective lenses prefer glasses rather than contacts. Either way there is some danger to the eyes in the event of a crash. A common injury in road accidents consists of lacerations of the face and eyes from broken lenses. Plastic lenses are less prone to breakage, but do not entirely eliminate the risk. Contact lenses can slip during a race, which will most likely cause the driver to stop. They provide an additional danger of coming out of place in a crash. If you wear contacts, notify emergency workers of this fact along with the other medical information on your helmet. If you need corrective lenses to drive, talk to your optometrist about the options.

Hand-to-eye coordination is also something that can be improved with practice. Years ago, I practiced with a stopwatch by trying to anticipate and stop the hands (it was the pre-digital era) at a specific position. When I began, my average reaction time was six to eight hundredths of a second. With practice, I was able to get to within four or five hundredths of a second. Now I practice by stopping videotapes and gas pumps at predetermined points. Practice is where you find it. Controlling a racing car demands precise timing. You will find these skills valuable in determining exactly the points on the track at which to begin braking or turning and combining the various actions required to get through a turn.

A human being is remarkably efficient at gathering information from his surroundings. While controlling the car, the driver accepts information through four major channels: vision, sound, motion, and kinesthetic sensations. What doctors call the vestibular apparatus is loosely defined as the balance organ or the inner ear. It is composed of three semicircular canals, oriented at 90 degrees to one another, that detect rotation or angular acceleration. When the head turns or a force (like a G force) acts on the fluid in the canals, they detect movement and send the appropriate message of rate and magnitude to the brain. These messages work in concert with the eye muscles, keeping the focus fixed on an object even if the object is moving through the field of vision while the field is changing.

The inner ear is also closely related to the muscles and joints through the nervous system. Angular and linear accelerations are compensated for by unconscious motor activity. This is most efficient when the accelerations are high intensity, short duration, and of alternating direction, the conditions encountered when driving a race car.

The kinesthetic sensations are more complex than just pressure-sensing cells below the skin. They also include sensors in the tendons and joints that signal pressure and position information about joint position and muscle stretch. As the car is jostled about, these sensitive receptors relay this information to the brain. The closer the connection between the car and driver, the more information the receptors will transmit. Accurate information regarding joint position and muscle strain are imperative to controlling the car through the steering wheel, pedals, and shift lever. Dr. Michael Henderson, a prominent British medical doctor and racing enthusiast explains, "The ability to react without conscious thought to what information is being transmitted, and to avoid disorientation from confusing and misleading sensations, is partly inborn and partly learnt. The 'inborn' proportion is a small part of what makes a 'natural' racing driver, but constant practice at racing speeds can so improve this processing of information that the determined driver can overcome what might have been a natural, neurological handicap."

Some people may be genetically more suited to becoming race drivers than others, but that does not mean that the ones who are slightly deficient in one area or another cannot improve those traits. Whether we talk of muscles or learned responses, they can all be made stronger, quicker, and more efficient by using them and stretching to reach full potential.

Chapter 2

Selecting Class and Car

To maximize enjoyment of the sport, the type of car you race should be suited to your budget, technical experience, understanding, and on-track experience.

The Type of Car You Race

Karts

Karts are universally misunderstood, underestimated by most, and overestimated by the rest. In reality, they are real competition cars, albeit miniature versions, with anything but miniature performance. They are divided into two basic groups: Sprints and Enduros. Sprint Karts are typically run on 1/4- to 1/2-mile road courses in heats of eight to ten laps each. They emphasize car control and getting up to speed quickly. In Enduro Kart racing, the driver lies down between two fuel tanks and runs on long tracks, in many cases, the same ones on which car races are run. The speeds are much higher than Sprint racing and the races last one-half to a full hour. Both are excellent training grounds for someone wanting to gain experience to move into car racing. Many people try Karting, decide it is just what they want, and stay. Those who have used Karting for experience, though, include most current pro drivers from Al Unser, Jr. to the late Aryton Senna.

Karting is good experience because Karts are light, have good power-to-weight ratios, low polar moments of inertia, low centers of gravity, locked rear axles, and fast steering ratios, and generally only have brakes on the rear wheels. This makes them *extremely* responsive with exhilarating performance and a generally oversteering nature. Although Karts will not run 180 MPH or do 0 to 100 MPH in 4 seconds, they will give you a rush you will not soon forget. If you can drive a Kart fast, you're halfway to driving a car fast.

As many classes of Karts are run as SCCA has classes, but one stands out as the most like cars in terms of performance and response: Shifter Karts. Most of these Karts are built like the Sprint variety, i.e., the driver sits upright, although many Enduro races are organized for them as well. They use either 125 cc or 250 cc two-stroke motocross engines with six-speed gearboxes, and have brakes on all four wheels just as God intended. They drive much the same as cars and the 125 cc versions have power-to-weight ratios similar to Formula Fords and run only slightly lower top speeds. Their response characteristics are more like Formula One cars. They have become very popular, with local Sprint races attracting 10 to 20 entries each week in some parts of the country and the premier Enduro events such as the annual street race in Elkhart, Indiana having over 100 entries. The larger displacement 250 cc division is not as popular, but offers substantially improved performance. After testing one, David Coulthard, current Formula One driver, commented that the acceleration was just a bit less than an F1 car, but the cornering ability was about the same and the brakes were better! Maricio Gugelmin was a contemporary of Senna's when they were both racing Karts. When ask what advice he would give a novice driver he replied, "Do as much Karting as possible."

Regardless of what type of Kart you might choose, it will provide good performance and outstanding responsiveness. Whatever happens in a Kart, happens *NOW!* Many car drivers keep a Kart to run between car races in order to stay sharp. Emmerson Fittipaldi is said to have run Karts between his Formula One and Indy Car careers. The misconception that exists among some car racers that Karting is beneath them is ill-founded. Experience in Karts will make you a better driver.

Although Shifter Karts will not reach really high speeds, they offer the low-speed acceleration of a Formula Atlantic and response characteristics of an F/1 car at an affordable price.

Oval Track Cars

Racing on ovals in America is by far more popular with fans and participants alike than is road racing. This is so because the spectators of an oval race can see close battles raging on the entire track, unlike on most road courses and all street circuits. Even local dirt track races pay the high-finishing drivers because the spectator base is better. These races are roughly equivalent to SCCA's club races, most of which are non-spectator events, and all of which pay no prize money. The oval track sanctioning bodies have kept the rules consistent with close competition and rarely change them, which helps to keep the costs lower and provides the close competition the fans love. The close racing and the generally enclosed nature of the tracks foster a lot of wall banging and fender bending, which the fans also seem to enjoy. The only bad thing that can be said about this facet of the sport is that, among the thousands of local oval tracks in the nation, few have consolidated their rules. This makes it exceedingly difficult for a racer to travel to another local track to race, let alone another part of the country. It also makes it difficult to describe the classes that are available to the oval track racer.

Closed Wheel Cars

Despite the lack of consolidated rules among oval tracks, similar classes exist in almost all parts of the country. They are usually different in name and have minor rules variations. The variations in rules may be as easy to deal with as requiring a certain type of tire, or as significant as requiring a reduced engine size. The closed wheel category usually begins with some semblance of Hot Stock. These cars are most often run on quarter-mile tracks of either dirt or asphalt, and are based on Monte Carlos, Chevelles, and Torinos. They use stock V-8 engines with 2-barrel carbs, street tires, and virtually stock suspension. These cars are the type in which Foyt and Andretti got their start, except the time frame was a few decades earlier and back then they were called Jalopies. The cars can be bought very cheaply (sometimes for as little as $400 to $600) and may be run almost as inexpensively, but, as the axiom goes, "You get what you pay for."

The next step up in performance is the Mini Stock class, which has shown a dramatic increase in popularity in recent years. Mini Stocks are Vegas, Pintos, Tercels, and similar cars that use somewhat modified power plants and offer a good balance among complexity, cost, and performance. Rarely do their engines exceed two liters, and minimum weights are in the 1600 to 1900 pound range, so neither acceleration nor top speed is likely to overwhelm the novice driver. A Mini Stock can be built for as little as $2000 or for as much as over $10,000.

Farther up the ladder is some form of a Limited Late Model. These cars are loosely based on Mustangs and Camaros. They use more highly modified V-8s and slightly more sophisticated suspension than the Hot Stocks. Weighing in at around 2000 pounds and producing 300 to 400 horsepower, they are a bit more serious than the Mini Stocks, although they carry price tags only a few thousand higher.

Up another rung are the Late Models. These are tube frame cars using highly modified V-8s and purpose built suspensions. Cars for all the oval track classes up to Late Model are traditionally built in the garage by enterprising racers using the "cut and try" method. Late Models are a bit more sophisticated, being built by commercial chassis builders using the same "cut and try" approach. A few of these cars have made their way to GT-1 grids (more about GT-1 later), but without much success. Some parts of the country run what they call Radical Late Models, which use the same concept, but all of the roll cage and top are removed except that enclosing the driver. This makes for better airflow on the passenger side of the car and makes them readily identifiable. Late Models and Radical Late Models are expensive cars using many specialized racing parts. They are expensive to run as well, but provide impressive performance.

Modifieds are quite popular in the Northeast, but it is difficult to classify them as open or closed wheel cars. They must be based on front engine, rear drive American sedans. In reality, the only stock part they utilize is a portion of the frame. They use stock appearing rear body panels and only a piece of sheet metal to resemble a hood up front, but use no front fenders. Their engines are big V-8s, 427s, 454s, etc. These are pretty serious race cars that make a great deal of torque and will make a driver's eyes get big when he opens the throttle. They are big and heavy, but with their torque they get around a short dirt or asphalt oval quite well. Expect to pay big money to buy a Modified.

Open Wheel Cars

Open wheel cars are usually considerably lighter than their closed wheel cousins, and offer a substantial increase in performance. Although race tracks in some parts of the country do run some really small open wheel oval track cars, such as those powered by Kart engines or 250 cc motorcycle power plants, we will begin looking at slightly larger ones that can properly be called cars. Of these, the first to consider is the Legends class, which has been growing like grass in all parts of the country. Legends are 5/8ths scale replicas of the Jalopies of the forties. Street tires are used as well as 1200 cc motorcycle engines with no modifications. The performance of Legends cars is brisk and the price tags are kept intentionally moderate.

A small step up are the Outlaw Minisprints. These cars are normally built by professional chassis builders, are safe and light, and use wings just like their bigger relatives. They are allowed the same 1200 cc four-stroke motorcycle engines of the Legends cars, but most organizations allow extensive modifications including electronic injection, ported heads, and methanol fuel. Where the clubs allow fully modified engines, they are making around 200 horsepower and propelling about 875 pounds including driver. This is enough to make any driver pay attention, regardless of his level of expertise. A used Mini Sprint can be purchased for $4,000 and up, and brand new state-of-the-art equipment will set its owner back about $15,000.

Midgets have been around since the wheel was invented. They have evolved over the decades into pretty sophisticated cars that work very well for one purpose only: to get around a short oval as fast as possible. About the same size as the Mini Sprints, Midgets weigh a little more, largely because of the four-cylinder automotive-based power plants they use. These engines are about 2.0 to 2.4 liters in displacement and crank out 300 to 350 horsepower. Needless to say, these cars are not for the faint of heart. Midgets do not use wings, so all the grip of the tires on the track surface must come mechanically. This performance comes at a price. Used Midgets can sometimes be purchased for around $10,000, and new equipment can easily top $40,000. Organizations such as USAC run some well-respected professional racing series for Midgets.

Sprint Cars are what most people think of when they hear the term, "open wheel dirt car." These cars are powered by American V-8s, usually based on the small block Chevy. Some tracks run what they call Limited Sprints, which are limited to 360 cubic inches and a four-barrel carburetor. Outlaw Sprints, however, use 410 cubic inch all-aluminum injected engines burning methanol. These are the short oval equivalent of Formula One cars. Don't even think of driving one of these without plenty of practice in something else first. Making 860 horsepower, weighing about 1500 pounds, and geared for short tracks, only one word truly describes the experience of driving an Outlaw Sprint: BRUTAL! A new World of Outlaws Sprinter will require $60 to $75 thousand to get on the track.

Saturday night short oval races are excellent experience for a driver. There he can learn to run close, plan race strategy, set up a car, and many of the other things every driver needs to know. The question of racing on dirt or pavement matters little to his progress on the learning curve. And because quarter-mile ovals do not allow really high speeds even in moderate-engined cars, a driver can practice his craft in comparative safety. Crashes are frequent, but not usually serious.

Years ago, the road to Indy began on these tracks. Today, though, the map has changed. Most of today's professional drivers have gained their experience in Road Racing cars. The exceptions, of course, are the Winston Cup drivers almost all of whom began on these short tracks. Sammy Swindell says, "I won the World of Outlaws title twice and that's enough for me. You could win it fifteen times and I don't think it would do a lot for you." A few Sprint car drivers have tried to make the switch from World of Outlaws competition to NASCAR and Indy with little success. Some feel that their failure to make the transition has less to do with moving from dirt to asphalt or from 140 MPH to over 230 than with the significant difference in car behavior. Sprint cars are extremely fast, but hardly nimble. Both Indy and Winston Cup cars behave quite differently from each other and from Sprint cars. Some background in similar cars is required for adaptation. Swindell and Kinser might have been better served by a season in Formula Atlantic or Indy Light before trying Indy Cars. Meanwhile, the road to Daytona and Talledega still begins on the nation's short tracks.

Road Racing Cars

It has been said that SCCA should have three classes: cars with roofs and fenders; cars with fenders, but no roofs; and cars with no roofs or fenders. While that oversimplification may cover the bases, a few more classes are needed to satisfy the many different tastes, racing philosophies, and budgets of racers. Road racing is broken down into five categories, but several variations exist for each of these.

Showroom Stock

In the beginning days of the Sports Car Club of America, a racer could drive his MGTD to a racetrack, tape up the headlights, and go racing. Showroom Stock (SS) is an attempt to return to those days. In theory, no modifications are allowed to the cars with the exception of safety features. Belts, fire extinguisher, and bolt-in roll cages are required to prepare a modern sports car for competition. The reality, though, is that it takes much more than these items to be competitive. The tire companies sell "shaved" street tires in racing compounds for use in SS. Too much tread depth causes the tread to squirm, which is not good for traction. Because these tires start life with less rubber, and because heat affects the performance of slicks and treaded tires alike, it takes fresh tires every weekend to run up front. Some Showroom Stock racers are reported to go to the expense of blueprinting engines to bring all specifications (and performance) right up to the maximum. Legality problems abound in the SS category. Many races have been won or lost in the tech shed over such items as a computer with the wrong part number stamped on it. Check the Protests/Appeals Column of the "Fastrack" section of *Sports Car Magazine* for any given month and you are likely to find that 50% of all the protests have to do with the SS classes.

The category is broken down into Showroom Stock A, B, C, and GT for various kinds of cars. Although the concept is sound, Showroom Stock doesn't completely live up to its heritage. On the plus side though, what other type of race car can be bank financed? I wonder if the insurance company knows how that little "fender bender" really happened on a "country road north of Atlanta."

GT Cars

For cars with roofs and fenders, there are several possibilities within the GT classes. GT is an acronym for Grand Touring, which is supposedly a production performance automobile, generally having two seats. Within SCCA's rule structure, cars built for these classes may be built from scratch using tube frames and aftermarket fiberglass bodies. The engine must be of the same type originally fitted to the car and in the same location, but can be extensively modified. Restrictions are placed on engine size, induction, fuel, number of gearbox speeds, weight, and aerodynamics, and minimal restrictions are placed on suspension type. GT classes range from the miniscule GT-5 Mini Coopers to the ground-pounding Camaros of GT-1. The GT-1 category is the club-racing version of the Trans Am pro series.

GT cars are considered by some racers to be safer than other types because the driver is contained entirely within the roll cage structure. It is true that this type of construction affords more protection than others, but three other factors should be considered. One is that it is more difficult to extricate an injured driver from a closed car. Another is that they are generally heavier cars than those of other classes, and therefore have more kinetic energy at speed to dissipate in the event of a crash. The third is that any class that uses fenders promotes rubbing door handles. This makes accidents more frequent. In practice, these factors are probably canceled out by the full roll cage, because it seems that no one type of car produces more injuries than any other. As a matter of fact, although injuries do occur, they are rare in all classes.

The smaller GT cars such as these GT-4s are easier for the novice to handle in terms of power, speed, and cost ...

... but the larger GT cars such as this Corvette driven by Lou Gigliotti are real fire breathers.

Another factor that may influence your decision as to which category of car to race is the ability to find sponsors. Full-bodied cars are more recognizable as a particular make and, as such, may attract dealers or manufacturers of aftermarket automotive parts more readily than other types. As you will see in Chapter 9, though, automotive sponsors are only a small portion of those available.

Regardless of category, the heavier or more powerful the car, the more expensive it will be to run. Beginning racers are usually on a strict budget. Unless yours rivals that of the Pentagon, you should stay away from GT-1. Engines for these cars can run as much as $25,000 each, and tires will cost almost $1,000 a set. Besides, 2900 pound, 180 MPH cars are not suited to the beginner. GT-2 cars are slightly more modest in both cost and performance. These include Toyota Celicas, Nissan 300ZXs, Porsche 911s, and the like. These cars can still exceed the budget and the required experience of the beginning driver. These are 150 MPH cars and can cost $40,000 to $50,000 to build. From the GT-3 to the GT-5 levels, the cost and performance potential are more realistic. In GT-3 you will find BMW 320i's, the 2.3 liter Mustang, and similar cars. In GT-4 are the Honda Civic and Toyota Corolla. GT-5 contains the Fiat 128, the 1300 VW, and the perennial Mini Cooper.

Production Cars

For open cars with fenders, the Production Cars are one choice. These are cars built from production sportscars, highly modified for racing, but not built from the ground up for the purpose, as the GT cars are. Production Cars are not allowed tube frames, but must have roll cages incorporated into the stock frame or unibody structure. The engine must remain in the stock location, but can be highly modified with specific restrictions on carburetion and valve size. All restrictions in the Production categories are more tightly controlled than in the GT classes. Limits are placed on weight, wheel size, and number of gearbox speeds. The suspension must be of the same type originally fitted to the car, i.e., independent, swing axle, solid axle, etc., but suspension parts may be substituted or fabricated.

The Production Cars have less engine displacement than the GT cars and are proportionately slower. The cars in E Production, Porsche Speedsters, 914s, and MGBs, have a speed potential of around 140 MPH. In F Production the car of choice is the MG Midget with the 1275 cc engine, which will run about 130 MPH. In G Production the Triumph Spitfire usually runs up front, and in H Production the Austin-Healy Sprite is the car to run with the 948 cc engine. Any of these cars in any Production class would be a good place to start.

Racers buy Production Cars because they are affordable. Many times a car can be purchased on the used market for $4,000 to $6,000. Unfortunately, as many who have bought cars in this price range have discovered, more money has to be spent to make them competitive in national races. The engines, gearboxes, and suspensions are then updated over the course of a few seasons, and when the car is finally running at the front of the field at the Runoffs™, an additional $10,000 or more has been spent on the updates. Production Cars provide a good way for the beginner to go racing if he knows from the outset that he will not be competitive without spending more.

SCCA has recently been adding other cars to the Production categories, such as the Miata in E and the VW Rabbit (a closed car) in F Production. As time goes by and some of the new cars are found to be either substantially faster or slower than their competition, performance adjustments will be made to the rules usually in the form of weight or carburetion requirements.

There has been a great deal of controversy in recent years about the SCCA trying to kill the Production classes. Whether this is fact or fiction, no one outside of Denver is really sure, but it seems to be founded on the fact that Production Cars are not popular in the Rocky Mountain Division. Whether Denver's influence is the reason, or the lack of local participation is something Denver perceives (incorrectly) on a national scale, HQ has given a lot of support to Production classes recently. The Comp Board has revised the Production Car Specifications to make the rules less ambiguous and, with the addition of the "new, improved IT cars" to the class, participation should increase. There are enough participants running Production Cars that, if SCCA really were trying to kill the category, such a stink would be stirred up that it would even be smelled in Denver. Then again, maybe it was.

Production Cars are generally more affordable than GT cars and provide a lot of diversity in design. Here we see a Mazda, an MG, a Lotus, and a Triumph.

Sports Racing Cars

Sports Racing Cars are supposed to be prototypes of real road going production sports cars. Anyone can see that in reality they are purpose built racing cars with fenders, but no roofs. They are always mid-engined and use fiberglass bodies. This category though, can be further broken down into spec classes and open classes. Let's deal with the spec classes first.

A few years ago, SCCA formed a subsidiary corporation called SCCA Enterprises whose purpose, supposedly, was to make a profit. That's when all the recent controversy began. The mission of SCCA Enterprises is to build cars for which SCCA Inc. makes the rules, homologates the cars, and conducts the races. Some said this was a conflict of interest. I don't know why. They began by instituting a class that was initially called Sports Renault and later became Spec Racer. Its rules are a formula for a mid-engined, tube frame, single seat race car with covered wheels using a stock Renault engine and specified Renault suspension parts. The theory behind this is to provide an affordable race car that provides close competition, is easy to maintain, and is inexpensive to race. The car actually fits the bill pretty well with some minor exceptions. Spec Racer racers will tell you that not all spec engines are created equal. There are engines and there are *engines*. *Engines* cost more. Tires are a big expense for cars that run slicks and the street tires of the Spec Racer should solve that. In reality, though, a fresh set of shaved street tires is required for each weekend of racing to run up front. While most racing expenses in other classes seem to be in tires and engines, the Spec Racer's big costs seem to be in fiberglass. The unwritten rule is that if you don't have tire marks on the body sides and fiberglass flapping loose on the nose, you

The SCCA Spec Racers are easy to drive and a driver will always find someone to race with.

just aren't trying very hard! At the June Sprints™ at Road America in 1990, the flagman put out the furled black flag to the *entire field* of Spec Racers! To my knowledge this had never been done before. These cars will teach you how to mix it up in close company. Recently a Ford engine has been added to the Spec Racer lineup and, of course, had to be separated from the Renault engined field, so another class was formed. A new pro racing series is even run for the new class.

SCCA's next foray into building cars was the Shelby Can Am. This is a larger version of the Spec Racer using a Dodge V-6 engine that is somewhat modified and makes gobs more torque and horsepower than the Renault's 80 BHP. As in Spec Racer, there are engines and there are *ENGINES*. Although the Can Am's are pretty crude as real race cars go, they will go pretty well. Their top speed is in the 160 MPH range.

Assuming a driver already knows how to *drive* a car and has good car control, the spec classes are good for teaching him how to *race*. Since everyone has exactly the same equipment, the cars are very closely matched and the driver will always have someone with whom to duel wheel-to-wheel. And they make a very good pretense of saving racing capital.

The non-spec sports racing classes include C and D Sports Racing and Sports 2000. D Sports is a technically innovative class in which the do-it-yourself car builder finds a lot of latitude for experimentation. The engines are either highly modified 1000 cc motorcycle engines, multi-cylinder two strokes, or 1300 cc automotive engines. Since the minimum weight with driver is only 900 to 1000 pounds, depending on configuration, they come alive when you light a fire under them. It is unfortunate that the engineering that goes into the best of the D Sports racers cannot be used in them all. The good ones are really good. Since many of the cars are built by the cut and try method the others tend to be really bad.

C Sports Racing cars are usually older Formula Atlantics or Formula Super Vees that have been converted to closed wheel configuration by changing fiberglass. Since they are based on monocoque tubs with sophisticated suspension, brakes, gearboxes, and engines, they are a tad on the expensive side to race. With a few notable exceptions, these cars have tended to run down a bit over the years, and we find the same situation as with the D Sports cars, either really good or really bad. A good one, however, will make 230 BHP and, in a car weighing about 1200 pounds, this is good enough for around 160 MPH and fairly breathtaking acceleration.

Sports 2000 cars are closed wheel versions of Formula Continentals. Rather than being of tube frame construction, they use the more expensive monocoque chassis, and are more expensive to buy than their open wheel cousins. Since they use the same drivetrain, though, the operating expenses are roughly equivalent. The engines are blueprinted Ford SOHC 2 liter units and are highly understressed, so engine maintenance costs are quite low. A used engine can be obtained for $3,000 or $4,000, and engines can be rebuilt in your garage for a few hundred dollars. If you choose to use a professional engine builder, expect to purchase one for around $12,000, and freshen it at five or six race intervals if you are running club races, or at one to two race intervals

Track Time Generates Success

When I was young, my eagerness usually got the best of me. Anytime I discovered an opportunity to drive a race car, I was strapped in before anyone could ask, "You got a license to drive that?" Because of my eagerness, I sometimes drove cars that were somewhat less than safe, and on other occasions, I got into cars that I had no business driving due to my inexperience. Although I was lucky not to crash in those cars in which the belts were held in by 1/4-inch bolts, it didn't take me too long to recognize the importance of proper safety procedures. In reality, every driver endures a couple of "near misses" in every session on the track. I realized that it was only a matter of time before I would need those belts.

I realized something else, too. When I got into a car that was too big or too fast for my level of experience, my awareness of what racing is all about expanded. The most graphic example is the year I had the opportunity to drive an unlimited Sprint Car on pavement. My most recent experience had been a few sporadic races in a Formula Ford and a lot of Karting prior to that. None of it prepared me for a 700 HP, 1500 pound, ill-handling, fire-snorting monster. Although that was the only full season I ever did in a big inch oval track car, I jumped several steps up the learning curve and also learned lessons that I have since applied to road racing cars, which outwardly would appear vastly different. An example is the selection of proper "stagger" across the rear end (discussed more fully in Chapter 8). When I learned how important stagger is to turning left in an oval track car, I realized how important zero stagger is to a road racing car. Now that Formula 2000 cars are running a great deal on ovals, that knowledge has become an "unfair advantage" which I can use against the drivers/engineers who have little oval track experience.

The same can be said for a driver's on-track experiences. I never became as proficient as I wanted driving that Sprint Car, but it did give me a taste of torque and acceleration like only Top Fuel Dragster drivers knew at that time, or so I thought. I realized that I must learn to control those high forces at the limit of adhesion, as I had done with my Formula Ford at lower levels of both lateral and longitudinal acceleration. I learned, too, that the Sprint Car had peculiarities in the handling department that I would have to learn to deal with if I could not change the car's handling characteristics in the pits. I learned to drive around some of its problems. This knowledge turned out to be another "unfair advantage" when I got into other cars.

if you are pro racing, for about $2,000 or $3,000 each time. Sports 2000s are pretty sophisticated race cars. The biggest costs involved are tires at around $700 per set, and, if you are prone to hit things, fiberglass and suspension parts replacement. These specially made parts can be quite expensive. Expect a pretty brisk performance level from these cars. Their 160 BHP will push them along at about 140 MPH.

Every car has its own personality. Some work beautifully when driven at 90% and turn into complete bitches at 95%. Others snarl and bark below 95% and above that become wonderful, civilized machines that are pleasures to drive. The engineer's task (that's you if you do not have an engineer on your team) is to make the car as easy to drive as possible. In this way, the driver has confidence that the car will do what he tells it and he is prone to push it harder. That's what makes the car/driver combination fast, but it is not likely that every car you drive will be so easy. For those occasions when you are driving a car that pushes on corner entry, or oversteers at the slightest movement of the wheel, or even just doesn't give you much information about what the front wheels are doing, you must know how to compensate for these problems. Experience in a number of cars teaches this. They need not have the engine in the same end you are accustomed to; they need not have a roof; and they need not even run on the same type of track with which you are familiar. Drive everything you can get your hands on to find out how it behaves, how it feels, and how to deal with its own peculiar problems.

An additional benefit of begging for any ride you can get is that you can rack up a great deal of seat time rather early in your career. Since driving a race car well is an acquired skill, the more practice you are able to get, the better you will be as a driver. No one can study the multitude of actions and timing of those actions so perfectly that his first lap on the track is fast—or his tenth or hundredth. It takes practice.

Pro teams recognize this and do a great deal of testing for reasons other than setting up the car and checking fuel mileage. The drivers need practice and they need it regularly to stay in top form. Mario Andretti says that a lay off of only three weeks means his first session on the track is used only to adjust his mental processes to racing speeds. In that first session, things are whizzing by faster than the mind is accustomed to seeing them and time is wasted in processing information. This results in poor judgement and mistakes. If this is true for a driver of Andretti's caliber and experience, who drives more in one race meeting than the average club racer does in a season, how much more important should it be to that club racer to get track time? Practice will never make a club driver perfect, but it certainly will make him a better driver, and the more practice and the more frequent that practice, the better. Drive everything you can—even if it means wearing a "Will Give Up Food to Race" tee shirt!

Of all of these, Spec Racer and Sports 2000 are good classes in which to begin. The others will be either too fast for a beginning driver or too expensive to run. You may find it difficult to find a good car in C or D Sports, and you will learn more quickly and with less frustration in a good car. Spec Racers tend to be pretty slow, but with a lot of competition. Sports 2000s are a lot faster and put more emphasis on chassis setup.

Sports 2000s are sophisticated cars requiring a pretty good driver with fairly deep pockets.

Formula Cars

Formula Cars get their moniker from the set of rules or *formula* governing their construction. Actually, all cars could be called Formula Cars because there is a set of rules for cars of all types. Formula Cars have come to be known, though, as open-wheel single seaters, in other words, cars without fenders or roofs. This category is for the purist, since there are no parts on the car that are not necessary for competition. Fenders, lights, windshields, passenger seats, etc., have all been discarded. In SCCA, Formula Cars range from the modest Formula Vees and F/500s to the very immodest Formula Atlantic.

If you like Karts, Formula 500 may be just the thing. These cars are powered by two-cylinder two-stroke 500 cc engines in which modifications are strictly prohibited. Rather than a transmission, a belt drive is used with a variable-ratio sheave. They are of tube frame construction and use 10-inch diameter wheels. Driving an F/500 is more akin to driving a Kart than a car. There is no clutch or shifter. At only 750 pounds, though, they go reasonably well. On some tracks they are a bit slower than the Formula Vees, but at other tracks F/500s excel.

Formula Vee (F/V) is a class for tube frame cars using many stock or almost stock VW components. The wheels, front suspension, steering, brakes, engine, and transaxle must all meet these specifications. Slicks are allowed, but are very narrow and do not provide a great deal of grip.

The engines of the very best Vees make only about 60 BHP, but the cars are quite light. Even so, the performance is less than exhilarating. The way to drive an F/V fast is to avoid getting sideways and scrubbing off speed. Because of this, F/Vs are good training for precision driving, but not so good for learning to handle horsepower. Formula Vee is a very popular class, one of the club's largest because of the low cost of initial purchase and racing costs, and because they are much easier to drive than some of the higher-powered cars.

Formula Ford (F/F) is another of SCCA's most popular classes, although the fields have been thinning since the mid-1980s. There have been a number of thresholds in F/F development over the years, but the last one occurred in 1983 with the introduction of the Swift. It was a quantum leap in performance. These days, Swifts dominate the front of the field in FF.

The 1600 cc engine makes about 115 BHP, and the cars weigh 1100 pounds with driver, which gives them a power-to-weight ratio of about 10 to 1, roughly the same as that of the Camaros and Mustangs of the muscle car era. Transmissions are four-speed racing transaxles, but lack the limited slip differentials of *real* race cars. They are tube frame cars with fiberglass bodies, and are required to use only 5 1/2-inch wide wheels. The tires, though, are unlimited, so the tire companies make tires with cantilevered sidewalls, which puts an eight inch wide tire on the 5 1/2 rim. Formula Fords are relatively easy to drive, but, like the F/Vs, don't like to be driven sideways.

Formula Continental (Formula 2000 in pro racing) uses the same drive train as the Sports 2000s and are very sophisticated, requiring understanding of vehicle dynamics for proper setup and a driver who really understands the meaning of sensitivity. Ryan Hampton is shown here on his way to a series championship second place in the 1998 F/2000 season.

The downside is that they have gotten a little pricey: A used Swift will go for $15,000 to $17,000. The market was saturated in the 1980s, so very few new cars are sold now. Formula Ford has been an excellent class since its inception, and has been a stepping stone for almost all of today's top drivers.

Formula Continental (F/C) has gained wide appeal in the last few years. The F/C is basically an F/F with the 2 liter Ford SOHC engine, wider wheels, and wings. Like that of its little brother, the engine is almost stock, and, as such, is relatively trouble-free and inexpensive to run. Operating expenses are almost identical to those for the F/F, and one competitor was heard to say, "It's the cheapest thing you can get with wings." Still, expect to pay from $18,000 and up. The performance of the F/C is pretty brisk. F/Cs are the open-wheel equivalents of Sports 2000s. The Ford engine will push the car along at about 145 MPH, and the wings allow it to corner at about 2.5 lateral Gs, enough to give you a real pain in the neck the day after a race. SCCA has a highly competitive pro class for F/Cs, which is thought to be a training ground for tomorrow's Indy Car and F/1 Car drivers.

Formula Atlantic is one of the club's fastest classes. Rather than using the tube frames of the other Formula Cars, Atlantics use composite monocoque tubs. They also use either 1600 cc Toyota or Cosworth 16 valve engines making 245 and 230 BHP, respectively. Five-speed gearboxes are equipped with limited slip differentials. The wheels are 10-inch wide front and 14-inch wide rear, and the large slicks on them provide massive amounts of grip. To take advantage of all that rubber, the cars are fitted with serious brakes. These are definitely not cars for beginners. They will run 170 MPH plus, and put a lot of drag racing cars to shame. This performance comes at a price, however. To buy a new car, expect to pay from $80,000 and up for a rolling chassis, and another $15,000 to $25,000 for an engine. A highly respected professional class exists for these cars too, and some find it silly to run them in amateur competition. SCCA has recently sold the pro Formula Atlantic series to CART, who is using it as a stepping stone to train future stars. Unfortunately, they are turning it into another spec class, as is their Lights division, by allowing nothing but Reynard chassis and Toyota engines.

We have not dealt with Regional-class cars and there is a good reason. If you intend to race for awhile, you will very quickly qualify for a National License. If you have selected a car that is only eligible for Regional competition, you will be doomed to run only Regional races until you can get another car. This means that you will only be able to drive for a couple of sessions on Saturday, therefore, you will not get as much opportunity to practice your skills. Some Regional drivers prefer this because they feel that they can run their races at a lower level of financial commitment and still be competitive. While this may be true, Karts may be a better choice for a beginning driver on a budget. For the purchase price of an I/T car, a Shifter Kart could be bought and would deliver much greater performance and learning potential. Since they are cheaper to run, the driver can get more seat time in a Kart, too, which is a big plus. One SCCA racer who had recently discovered Karts commented, "This is great. For what it takes to rebuild my Production Car engine, I can *buy* a Kart engine!"

CLASS COMPARISON

SHOWROOM STOCK

Non-modified street car with only safety equipment added

	Car Make	Initial Cost	Racing Cost	Top Speed (mph)
SSA	Supra, 300ZX, etc.	$18,000–$25,000	Low	140
SSB	Exclusively Miata	$15,000–$18,000	Low	130
SSC	Miata and Dodge Neon	$15,000–$18,000	Low	125
SSGT	Camaro	$18,000–$25,000	Low	140

GT

Purpose built tube frame cars with fiberglass bodies, original engine type highly modified

	Car Make	Initial Cost	Racing Cost	Top Speed (mph)
GT-1	Camaro, Mustang, etc.	$40,000 used to $100,000+ new	High	180
GT-2	Celica, RX-7, etc.	$20,000 used to $50,000+ new	Moderate to high	155
GT-3	Paseo, CRX, etc.	$6,000 used to $30,000+ new	Moderate	135
GT-4	Tercel, Sentra, etc.	$6,000 used to $30,000 new	Low to moderate	125
GT-5	Starlets, Sentra, Mini Cooper	$6,000 used to $30,000 new	Low to moderate	120

PRODUCTION

Production based with cage added to stock frame/unibody, original engine type highly modified

	Car Make	Initial Cost	Racing Cost	Top Speed (mph)
E Production	MGB, 914, etc.	$6,000 used to $25,000 new	Low to moderate	140
F Production	MG Midget, Spitfire	$4,000 used to $20,000 new	Low to moderate	130
G Production	Triumph Spitfire	$4,000 used to $20,000 new	Low to moderate	120
H Production	AH Sprite	$4,000 used to $20,000 new	Low to moderate	115

SPORTS RACING

Open cockpit, closed wheel purpose built mid-engined cars

	Car Make	Initial Cost	Racing Cost	Top Speed (mph)
Spec Racer	Spec Racer	$9,000 used to $14,000 new	Low	115
Spec Racer Ford	Spec Racer Ford	$10,000 used to $15,000 new	Low	115
Shelby Can Am	Shelby Can Am	$20,000 to $30,000 used	Moderate	155
Sports 2000	Swift, Lola, etc.	$20,000 used to $50,000 new	Moderate	140
D Sports	Mostly homebuilts	$6,000 used to $30,000 new	Moderate	140
C Sports	Ralt, Tracer	$15,000 used to $50,000 new	Moderate	160

FORMULA

Open cockpit, open wheel single seat purpose built mid-engined cars

	Car Make	Initial Cost	Racing Cost	Top Speed (mph)
F/500	Raptor, Red Devil, etc.	$3,000 used to $10,000 new	Low	115
F/Vee	Caracal, Mysterion, etc.	$3,000 used to $10,000 new	Low	115
F/Ford	Swift, Van Diemen, etc.	$15,000 used to $25,000 new	Low to moderate	140
F/Continental	Swift, Van Diemen, etc.	$15,000 used to $40,000 new	Low to moderate	145
F/Atlantic	Swift, Ralt, etc.	$20,000 used to $100,000 new	Moderate to high	165

Having examined the categories and classes of club racing, we need now to consider the factors that are important to a beginning driver. Foremost is a novice's need to get time on the track. Learning to race has been compared to playing a musical instrument. In the analogy though, the instrument is very complex, requiring a great deal of tinkering and tuning at home. When the instrument is finally ready to be played, this can only be done by taking it to a stage and practicing in front of an audience. Drivers in SCCA commonly make their way up the learning curve very slowly, in large part because they can only practice their driving a few times a season, and even then, for only a hour or two each time. This is where oval track racers have a big advantage in being able to practice every Saturday night. It is extremely important for a driver to get as much seat time as possible early in his career. For this reason, he should stay away from cars that are technically complex. Complex cars are expensive to run, which reduces track time for those who are on limited budgets, and they break more frequently, too.

Cars that produce a lot of power should be avoided as well. Actually, these cars are almost always the more complex ones, although they do not have a monopoly on technical sophistication. The high-powered cars are the ones everyone crowds to the fence to watch, but very few drivers have ever started in a fast car and been immediately successful. It takes a lot of seat time to learn how to handle that kind of speed and power. You will learn much more rapidly by starting in a lower powered car and moving up as you are ready to handle a bigger challenge. In addition to providing more enjoyment, your chances of avoiding an ambulance ride are better, too.

If a driver's goal is to bring home a trophy from each event so that his coworkers and the neighborhood kids think he's cool, then competition is detrimental to his purpose. In such a case, the driver should choose a class in which he has little or no competition. Everyone else, though, is likely to get more from racing by having someone to race with. If you are a member of this group, choose a popular class. In most Divisions this means F/F, F/C, S/2, F/V, or Spec Racer. Participation varies in different parts of the country, though, so check the competition in the division you plan to run. Production cars are quite popular in the Southeast, and participation in GT-4 seems strong in the Northeast.

If your goal is to get into professional racing, you would do well to consider a purpose built race car. Although, somewhere along the line, you'll need to run a car that is similar to the one that is your ultimate goal, it need not be your first car. If you are headed toward Trans Am or Winston Cup racing, a front-engined production-based car would be good experience. If you lean more toward Indy or F/1, Formula Ford is a good place to start. All along the way, valuable experience can be gained in many car types, and many young professional drivers make the most of this diverse education. Of the current group of Indy Car drivers, Gordon comes from off road and desert racing; Pruett started in Karts before making a name in Trans Am; Vasser ran F/F and Formula Atlantic; and Unser came up through Karts, Sprint Cars, and Can Am. Your first car is a serious decision, but not a permanent one. If you find it does not suit your purpose, you can sell it and buy another. However, if you spend some time researching the question, it may save you some headaches and a few dollars in finding the right category, class, and car. SCCA has many classes for many different philosophies of competition and different budgets.

Regardless of which class you choose, make sure you can afford it. It is much too common to pick a class for its appeal, with not enough regard for the financial commitment it will take to be competitive. I have known several racers over the years who bought or built the finest car they could, but then could not afford to race. Even worse, though, is the racer who sacrifices his family's well being to race. To be successful, the class you pick must be one that matches your budget. Before choosing a class, talk to other racers running the class in which you are interested. Keep in mind that racers running at different levels will have different budgets for the same class. Try to get honest answers to questions of cost. When you think you have a good estimate of the budget required to run at the level of competition you want, multiply it by 150%. Costs are *always* underestimated.

Just as often, a beginner overestimates his own abilities. We, as spectators, are so used to seeing Formula One, NASCAR, and Indy Cars running 170 to 240 MPH that we allow ourselves to be lulled into thinking it is easy. For a driver not used to the speed, 150 MPH is *very* fast! Under the right circumstances, much lower speeds can be very fast too, such as a downhill, off camber turn with bumpy pavement, and trees whispering in your ear. It can be very difficult to keep your right foot down even in a slow car in a situation such as this. When selecting a class to run, don't fool yourself into thinking that putting on a driver's suit automatically changes your name to Jacques and makes you able to handle his kind of car. It won't.

Spec Cars

Several professional series have been established in recent years using identical cars. The rules are quite restrictive in these classes allowing few, if any, modifications to the car or drive train. The intent of these classes is twofold. First, by using identical cars, it is the driver's ability rather than his budget or his innovative talents that propel him to the front of the grid. Second, by eliminating the innovations that racers inevitably want to try on their cars, the cost is kept to reasonable levels. The Barber Dodge, Spec Racer Pro, Formula Mazda, and even Indy Lights score well on each of these scales.

As with all other classes, there are downsides to the spec classes, too. Although the cars are supposed to be equally prepared, the reality is that favored drivers in a series sometimes seem to receive better cars. To become a favored driver, you must spend some money with the series promoter on schools or car rentals and buddy up with the right people. When you are a member of the right clique, you have a better chance of winning. Another downfall is that, although the spec classes are very good for minimizing costs while leveling the playing field, participants are able to learn little about chassis setup. As you will see in Chapter 8, learning these skills is crucial to a driver's development. Some drivers have run a spec series for several years and been quite successful, only to fail when changing to another class, where adjustments and modifications are the rule. As stepping stones, however, these classes work quite well.

Spec cars like these Formula Mazdas are generally inexpensive to run and provide close competition, but lack the chassis adjustments of more technically advanced cars, and teach little about chassis setup.

Choosing a Car

Choosing the class you want to run is only the first step in getting to the track. Now you must choose the type of car you should run. This begins not by looking at a group of cars for sale, but in the rule book. Don't be tempted to race a particular car because your friend has one for sale, or because you have found one that looks really hot. Choose the type of car that seems to have the best performance advantages allowed by the rules. The goal is to win races. We will get to cars built by racing car constructors like Lola, Van Diemen, or Swift in a moment. For now, look at the specification pages in the appropriate rule book. We will use SCCA Club Racing's F Production as an example. We want to pick a type of car that will have as many good characteristics, and as few negative ones, as possible.

Several vastly different cars are legal for competition in F/P. Each has different characteristics and advantages, as well as disadvantages. Generally, you should avoid heavy cars. Cars with small engines should also be avoided, too, unless they have a substantial break on weight. Adequate carburetion is a necessity, and the type of suspension required is also a pivotal factor.

The PCS (Production Car Specifications) list all cars eligible for competition in F/P. Let's examine some of their characteristics. MG Midgets are the class of the field. They win the majority of the races and championships. Always look first at what others are using to win. Because they are winning with a particular car does not mean it has all the advantages, but it is a good place to start. A Midget must weigh in at 1584 pounds with a 1275 cc engine and can use two 1 1/2-inch carbs. The 1500 cc engine is also eligible, but the car must weigh 1700 pounds and can only use one 1 1/2-inch carb, a very bad deal. Of these two, the 1275 cc Midget is the obvious choice.

The Triumph Spitfire is another choice in F/P, but is not traditionally as popular as the Midget. Let's see why. An MK-3 Spitfire must use a 1296 cc engine and can also use two 1 1/2-inch carbs, and must weigh 1600 pounds. These requirements are not too much different from those of the Midget. The big difference is in the rear suspension. The Midget uses a solid rear axle, while the Spitfire must use an independent, but of the "swing axle" type. The swing axle has severe disadvantages in that its roll center is (normally) much too high and it has too much camber change. (See Chapter 8 for a discussion of the importance of these terms.) Let's just say for now that the Spitfire has significant disadvantages in rear suspension design. If solutions for these rear suspension-related problems could be found, the Spitfire might be competitive (or even faster) than the Midgets. Until then, let's keep looking for advantages.

The Fiat X-19 uses a 1500 cc engine, is of a mid-engine design, uses a five-speed transmission (rather than a four-speed like either the Midget or the Spitfire), and uses a longer wheelbase for more stability. The Fiat has won the Runoffs™ in recent years, but must weigh 1795 pounds and use only one 34 mm carb. These are *BIG* disadvantages. The Comp Board at SCCA actually does a pretty good job of keeping things relatively equal.

We are looking for inequalities, though, that will give us an advantage. The Lotus 7 is only allowed a 997 cc engine and two 1 1/4-inch carbs, but its bore and stroke are vastly over square. This gives it more RPM potential, which will help to make up for the power lost to its small displacement. The Lotus 7 has a longer wheelbase than any of the cars discussed so far. Its big advantage, though, is its 1025-pound minimum weight. A savings of almost 600 pounds will make up for a great many other disadvantages. The Lotus may be the sleeper in the F Production class.

It seems a bit ironic that what we call Production Cars are really built one at a time, and each is somewhat different, while Formula and Sports Racing cars are mass produced and delivered identical. If you choose to run one of these factory-built cars, some of the same decisions required for other types must be made. Unlike other categories, though, the rules and specifications are identical for these cars so you might think there would be little difference in performance. The Swift DB-6, for quite some time, was dominant in Formula Continental and F/2000. Recently, however, the newest offerings from Van Diemen and Tatus have been winning a lot of races. These cars must use identical engines, so any difference on the track will come from suspension or aerodynamic influences. To make a decision on which car may give you the best chance of winning, you

Racing Classes Explained—Almost

If some of SCCA's classes are difficult to understand, it is not without reason. It has taken decades of dedication and concerted effort to bring the class structure to its present stage of confusion. Europe in the '60s had a stepping stone system which worked quite well to train pro-caliber drivers using Formulae 1, 2, and 3 (F/1, F/2, and F/3). Formula 3 was for small displacement, single-seater race cars, which were not too expensive to run. When a driver graduated from F/3, he went into Formula 2 where he could race against seasoned professionals, some of them F/1 drivers. When he was ready, or when he could get a ride, whichever came first, he moved into Grand Prix racing.

On this side of the pond, SCCA adopted a similar system in club racing using Formulae A, B, and C (F/A, F/B, and F/C). Formula A was for cars using stock block American V8s of five-liter displacement. Internationally, this class was known as Formula 5000. Formula B included cars a bit smaller using 1600 cc Lotus Twin Cam engines producing about 170 horsepower. Formula C cars were similar, but used 1100 cc four cylinder engines. For the 1973 season, SCCA changed the formula for F/B, allowing the Cosworth BDD 1600 cc four valve engine and renaming the class Formula Atlantic, presumably because the cars came from across that ocean or perhaps only because they were still trying to copy what was happening on the continent.

The original Can-Am series consisting of two-seater open-cockpit sports prototypes with unlimited engine size had been killed by spiraling costs when Porsche came into the fray with the Donohue-developed twin turbocharged 917-30K. In 1975, SCCA attempted to bring back their premier series by telling all the F/5000 competitors that they must put full bodies on their cars and call them Can-Am cars. Although this series was successfully run from 1976 to 1984, it never really had the thunder of the original series. In the late '80s, SCCA again tried to revive the Can-Am, this time using its experience constructing its own Spec Racers by building another spec car, the Shelby Can-Am. Using Dodge V-6 engines and having help from Carroll Shelby (formerly with Ford and originator of the Cobra), A pro class was run for them but has now folded and club racing is now the only place racers can compete with a Shelby Can-Am. For the (hopefully) final incarnation of Can-Am, SCCA has started a class with rules identical to PSR's Word Sports Car (WSC) class in an obvious attempt to steal PSR's competitors. These are serious cars with participation from major manufacturers such as BMW and Ferrari. It would be wonderful to see Can-Am return to its original splendor and it now has the best chance of doing so it has had in ten years.

During the '60s, there were also classes for closed wheel, open-cockpit two-seaters in club racing called A, B, C, and D Sports Racing (A/SR, B/SR, C/SR, and D/SR). A/SR was the amateur version of the original Can-Am classification, but used displacement limits. Looking back now, that seems to have been a very special time. You could go to a Regional/National and see Lola T-70s and T-165s, McLaren Mk-6s and Mk-8s, and other really high-powered ground pounders. B/SR was the province of smaller cars with two-liter versions of the Cosworth BD engine. C and D/SR were smaller yet and are the only of those classes to survive to the present day.

Formula V has been a low-performance, entry-level class since the early '60s. In the early '70s, SCCA instituted another class for cars using similar but highly modified VW engines, called Formula Super Vee. However, the engines proved to be grenades which came from the engine builders with the pin already pulled, so Volkswagen of America (VW of A) got involved in 1978 to promote the Rabbit and SCCA allowed the 1800 cc Rabbit engine. Around 1988, VW of A no longer wished to promote the car and pulled its money out of the Super Vee series. At this point, SCCA folded and left competitors no place to run their cars. Many of these relatively expensive cars wound up with full bodies in C Sports Racing and some found their way into Solo racing.

Meanwhile, Formula Ford had a bigger brother in Europe called Formula Super Ford, which used a variation of the two-liter Pinto engine and included wings. It became popular in all corners of the world before it caught on here, but eventually SCCA recognized it as Formula Continental or F/C. The original F/C had died out years earlier after F/B became Formula Atlantic or F/A. This was about the same time that F/A became A/SR or Can-Am. Confused? Good. You're normal. F/C is known in pro racing circles around the world now as Formula 2000.

We also had A and B Sedan classes, which were for American muscle cars with big block and small block engines respectively. They were allowed slightly more chassis modifications than their A and B Production cousins. There were also classes for C through H Production. C Production was dominated for a time by the Nissan, née Datsun, Z cars, and Paul Newman won a National Championship at Road Atlanta in one of these. In the early 1980s, SCCA combined A and B Sedan with A, B, and C Production to form GT-1, 2, and 3. GT-4 and 5 were newly formed classes. D Production was combined with E Production, and that is the reason that the Production classes now start with E and go through H.

Those are the National classes. We will not even go into the Regional classes such as Improved Touring or the new Sedan Classes. (Yes, Virginia, there is again an A Sedan class, but it is definitely not for Camaros with 427s!)

should know a little about the dynamics of race cars to compare them objectively. Carroll Smith's book, *Tune to Win,* is required reading on the subject, and Milliken's *Race Car Vehicle Dynamics* is the Bible.

Another important factor is driver comfort in these cars. In most other cars, the seat or wheel may be relocated to facilitate comfort. The small driver's compartment in many Formula Cars prevents this for average sized drivers. Although fast, the Van Diemen, for example, is very tight in the upper thigh area, and some drivers have complained about this. Make sure you can get comfortable in the type of car you choose.

Choosing the type of car you should run is not always easy. You are not looking for a car that looks fast, or even one that wins the majority of the races for other people. The object is to choose the type of car that will give *you* the best chance of winning races. Look at each car that is eligible for competition in the class you have chosen and evaluate its pros and cons objectively. Some people look upon a race car as an art object. This should be left to those running vintage cars. For the rest, a race car is a tool to be used by the driver to win races. If this is you, use the best tool you can.

Having chosen the type of car you wish to run, you must now look for a specific car. Almost every car type approved for competition has been built and run by enterprising racers at one time or another. The trick is to find the one that is the best prepared for the best price. As you will see, it is helpful if you know something about the specific type of car you are looking for. If you do not, take someone with you on your shopping trips who does. This includes cars from Formula and Sports Racing categories, of course, not just the GT, Showroom Stock, or Production Cars. The analysis can be broken down into the following categories:

Equipment. Let's say we have found a Midget for sale and it uses a 1275 cc engine. This is the type of car that wins many club races in FP. Most of the front-running Midgets use professionally built engines from a handful of builders. The one in this Midget seems to be a conglomeration of parts from several suppliers, none of which is very well known. The engine in this car, then, has not been dyno tested to determine torque, horsepower, or power band. It may be a good one, but it is impossible to tell.

The transmissions that come in the street Midgets are wide ratio gearboxes, as in most street cars, and in a racing situation, only third and fourth gear can be used. It is important, then, that a car have a box that has been modified for competition so that all four gears can be used. Let's say our example Midget's transmission has been fitted with Hewland type gears, which implies that proper ratios have been fitted.

When the Midget (and all Production Cars) was designed, the purpose was to cruise winding country roads on hard, skinny, tall tires while driving at what we would consider to be about three tenths of the limit. Consequently, the stock suspension that is still used on a number of Production Cars is not up to the task of dealing with wide, short slicks and lower ride heights while cornering at 1 1/2 lateral Gs. As a matter of fact, stock suspensions will not allow a car to corner that hard, thus limiting its speed through a corner, requiring more braking into it, and reducing the top speed reached on the following straight. A properly designed suspension will increase speeds through the turns by as much as six to eight mph! It is imperative that your car has well-designed suspension if you are to be competitive. Our example Midget uses a three-link rear suspension with coil over shocks, and is considered state-of-the-art. The front suspension, however, is basically stock with the addition of tubular shocks and stiffer springs and anti roll bar. This is hardly competitive equipment.

The seat in this particular car is a dune buggy type (see the Safety Equipment section of Chapter 5 for more information on seats), and the belts are faded, indicating significant age. The steering wheel is quite large and probably from some type of street car. Although it is adequate to steer the car, proper feel and control may require a smaller racing wheel. SCCA requires on-board fire extinguisher systems, and this car has only a hand-held extinguisher mounted in the driver's compartment, which means the car has not been run since this rule was initiated. Basically, all of the car's driver and safety equipment must be replaced.

Wheels are another important concern. Most Production Cars use one-piece cast-aluminum wheels that are quite heavy. Only occasionally will you find one that uses light, three-piece modular wheels because they are much more expensive, but worth it. This Midget has the normal heavier wheels.

Other items that warrant investigation are details such as whether the car uses the stock radiator or a more appropriate custom-built unit, the type of tachometer used, and whether or not the larger brake rotors and calipers that are allowed (and necessary) have been fitted.

To summarize, this Midget has a questionable engine, a good transmission, and needs a front suspension upgrade. The wheels are heavy, but since most everyone else's are too, at least this is not a severe disadvantage. The safety equipment should be replaced, and some of the smaller details such as the radiator may need to be updated, too. Expect to spend two to four thousand to bring the engine up to speed and another two to three thousand on suspension. All of the other parts may only add up to a thousand. Of course, if you are purchasing a Spec Racer, all equipment is *supposed* to be equal and none of this *should* be a concern.

When thinking of buying a car always ask to see its Vehicle Log Book. This is a running history of everything that has (officially) happened to the car. If it is involved in a major accident at a particular race, it will be noted, as will the severity of the damage. When an improper safety item is found at a tech inspection, it will be noted as well. Sometimes tech inspectors will let a questionable item slip through for a race, but will require it to be properly repaired before the next event. All of these notations in the Log Book will give you a clue as to how much trouble the owner has had with tech inspectors, and how diligent he has been in keeping the car up to minimum standards.

Condition. Regardless of the category of car you are considering, its condition will be an important buying consideration. Since the engine should be updated anyway, its condition is not too important. Any of the parts you intend to use, however, must receive serious consideration in terms of their condition. The wheels, for example, may be of a usable type, but if the rims are dented or they show signs of obvious repair, their condition will dictate replacement. Check the condition of all major parts carefully. The condition of all of the smaller parts and subsystems should be checked as well, for example, to make sure the radiator does not leak, the tach works, etc. As a general rule, though, whatever the condition of the large parts will also be that of the smaller ones. If the engine will not start, gears are rusty, and many rod ends need replacement, it is unlikely that you need to check the U-joint or CV joint condition. If you find problems with the major components, repair bills will be high regardless of the condition of the smaller items.

When checking these parts, some procedures may be helpful. Ask to hear the engine run. If the seller cannot comply with this request, use jumper cables from your own car, buy a couple of gallons of race gas, or whatever it takes to start it. Listen for unusual mechanical noises and the crackle of the exhaust note. Look out for either black or blue-gray exhaust smoke. Look for wobbling pulleys or any other anomaly. All of these will give you clues as to the engine's condition.

Mid-engined cars make it easy to check the condition of gearbox internal components, but the seller of a front- or rear-engined car may not let you pull the trans or rear end to take a look!

If the car you are considering is a mid-engined car with a competition gearbox, it can be easily disassembled to check the condition of gears, bearings, and dog rings. If the car is front-engined, though, it is doubtful the seller will want to pull the trans and open it up to let you inspect these items.

To check suspension bearings, jack the car up and try to rock the wheels vertically. This will indicate any slop in ball joints or rod ends on the uprights. Rocking them horizontally will show slop in steering joints. Any play in the wheel bearings will be seen in both of these tests. Each rod end or pivot bushing should be checked for play as well. If a rod end is loose, it will show it first on the axis with the bolt. Any slop whatsoever is cause for replacement. Teflon-coated rod ends begin tighter and last longer than the cheaper ones. On a Formula or Sports Racing car, total rod-end replacement may run as much as $1,500. Without disassembly, the only way to check the condition of wheel bearings is to check for play and rough spots while rotating the hub by hand.

Other parts should be inspected, also. Check the condition of engine belts, hoses, wiring, brake pads or shoes, distributor parts, U-joints or CV joints, etc. Expect to replace many of these items even on the best of cars.

Preparation. Condition and preparation usually go hand in hand. If a car looks neat and tidy, and the wires and hoses are neatly routed, grommets are used where they pass through panels, and there is evidence of much nicely-done safety wiring, it is quite likely the condition of the individual parts is excellent also. On the other hand, most club racers are on limited budgets and cannot afford to replace some parts before they are totally worn out. These racers will also tend to spend less time on the car than it deserves, maybe to keep the wife and kids happy, and do not perform many of the time-consuming tasks required to prepare a car nicely. These things include drilling bolt heads for safety wire, routing hoses in the engine compartment parallel to each other, and fabricating proper brackets for mounting components, rather than just drilling a hole in something to mount them. If a car has just won a National Championship or placed highly, it may well have had excellent preparation, but some of the parts may be tired. If it is touted as being a very good car because it won a championship a couple of seasons ago, but has been prepped by other people since, expect that it is not prepared as well as it was when it won. Most club racers are not fluent in racing preparation. For a complete course in preparation, read Carroll Smith's *Prepare to Win,* and commit it to memory.

Spares. Our example Midget includes among its spares three sets of wheels, two mounted with slicks and one with rains. The seller says that according to the car's Log Book it has not been run in two years. All of the slicks are "blue glazed" or have a slight-bluish cast to the rubber, especially in the upper sidewall area. Tires in this condition are very old and are totally worthless. The condition of the rain tires is usually the last thing the club racer worries about. Because of this, the rains are probably worse than the slicks. Two sets of wheels are Revolutions which are good racing wheels, but somewhat heavy, and the third set are street car "mags," which are *very* heavy. If the car had BBS or Jongbloed wheels, it would be worth more.

It also has three sets of gears and two dog rings in the spares box. The gears may have some wear, but are usable, but the dog rings have rounded corners and are worthless. Two bearings are included, but both have rough spots and are not usable. In with the spares is a front air dam which is cracked and worn on the bottom with missing Dzus fasteners. Also included is a stock transmission and clutch which were removed to install the race trans. Unless you have a street Midget, these parts are worthless. Two new crank fire pickups are in the spares box, indicating that this may be a trouble-prone item. Most sellers include a number of highly questionable parts in the spares to make the deal more attractive. Don't be fooled by them. Our Midget with its limited quantity of usable spares does not score highly in this category.

Price. The car's asking price should be evaluated against the above categories. In this case, it scores a six out of ten in the equipment included. Its condition and preparation may only be fives, and it scores only a three on spares. It could be transformed into a competitive car, one in which you could use those advantages you saw in the rule book, but it would take a great deal of work and money. If the asking price reflects the poor condition, it might be an acceptable basis from which to build a good car. In our example, expect to spend five to eight thousand dollars to make it a top-notch front-running car. Look at several similar cars and find out what the asking price of the best one would be. When you subtract the amount you will spend updating the example Midget, the figure should be close to its asking price. If you elect to take this route, be sure that you have the time, money, and energy required for the job. It will take more of these resources than you expect. Otherwise, keep looking. If you are fortunate enough to find a car that scores a perfect ten in each of the first four categories, you can strap yourself in, not spend a dime on it, and drive to a National Championship—provided you can drive. How to do that will be explained Part Two.

Building a Car

Purchasing an existing car is not the only option. To build a car takes a great deal of specialized knowledge, more time than you can imagine, and an absolutely incredible amount of money. If anyone tells you that a car can be built cheaper than it can be bought, it is a sure sign that he has never built one. If you decide to take this route, though, you can build in some competitive advantages, learn a lot, and enjoy having a complex assembly, one that you created yourself, actually be driven around the track at speed. Expect to spend a couple of years building your super car before you can enjoy driving it. If your goal is to be a successful driver, this road will be a time-consuming diversion.

Chapter 3

Preparation

Anything that can go wrong sooner or later will.

Fitting the Car to the Driver

One of the basic differences between competition driving and driving a street car is the greater sensitivity with which you can and must drive a race car. Most street-driven cars, even performance cars, tend to isolate the driver from the road to provide a comfortable ride. This is done with thickly padded seats, power steering and brakes, and soft springs, and with suspension geometry designed with comfort as the first priority. In a race car, however, the seat is made from a hard material with little or no padding, brakes and steering are both mechanical and direct, springs are much stiffer, and the suspension is designed for nothing other than traction. All of these features connect the driver firmly to the actions of the car. In this rather harsh environment, it may be difficult to understand how the driver could be comfortable. But comfort in this setting does not mean being isolated from the bumps in the road and vibrations from the suspension and drivetrain. You need those things. Being comfortable in a race car means being seated in the proper position so that you can use the proper muscles effectively to operate the controls. Any movements that are awkward will make precise car control difficult.

In the '60s, drivers preferred a seating position that straightened the arms when hands were on the wheel. Since then, we have realized that this position uses the deltoid muscles in the shoulders and upper arms for steering, and that these are not the best ones for the job. Seating position should be close enough to the wheel that the elbows are slightly bent, which uses the brachialis, triceps, and brachioradialis muscle groups in the forearm and just above the elbow. The knees also should be slightly bent so that the gastrocnemius and biceps femoris muscles in the calf and thigh can be used to pressure the pedals, rather than the muscles of the ankles, which must be used when the legs are straight. In most cars, this position is upright to semi-reclining. Most passenger cars and modern racing cars are built to accommodate this position and most racing seat manufacturers produce seats to fit this attitude, too. The lie-down position used in many cars of the '60s and early '70s is not adequate for proper car control, and fatigue sets in much earlier when driving in this position.

Unless you are a fighter pilot, both lateral and longitudinal forces are likely to be much higher in a race car than those to which your body is accustomed. To avoid using muscles unnecessarily to hold yourself in the proper position, it is imperative to be well-braced in the car. Having almost enough lateral support is bad enough, but it can be worse. I have sometimes tested cars that were not set up for me. The seats in these cars were frequently too large and I was, many times, too far from the wheel and pedals, requiring bundles of shop towels, extra Nomex, etc. to be used behind and beside me. These have rarely ever been in the right places to give adequate support for very long, and the result is my sliding around in the driver's compartment, and having to right myself after each turn. It should be apparent that these antics do little for car control, let alone sensitivity.

While driving a Ralt RT-5 Super Vee this way once, and using extra Nomex™ underwear as padding, something started flapping over my shoulder down the straight. Pulling it out, I found that it was a Nomex bottom. Since those things are expensive, I certainly didn't want to lose it, so I put it down between my legs for safe keeping. This didn't work well, however, as it kept flapping up and wrapping around my arms and the wheel. I pulled onto pit road, slowed down and threw it at the crew chief as I went by and pulled back onto the track. After the session was over he said, "I've been in this business a long time, but that's the first time I ever had a driver throw his underwear at me!"

To drive well, the hips and shoulders should be confined to prevent lateral movement. The belts should be tightened to hold the torso into the confines of the seat. (More information on the seat and belts can be found in the section on safety equipment in Chapter 5.) A car should be equipped with a dead pedal to the left of the clutch, which will also help brace the body in times when no braking or shifting is being performed, and aid in keeping the left leg from bouncing around.

The relationship between the throttle and brake pedal is important, too. They should be close enough to each other to make the foot's transition from one to the other quick and easy, but far enough apart to avoid hitting the brake when you only want the throttle. (The opposite is even worse!) Their heights should be adjusted so that when the brake is fully depressed it is even with the throttle. This will allow painless "heel and toeing."

If your car has other controls such as a brake bias control or anti roll bar adjuster, they should be placed where they can be found easily, without searching. It is sometimes necessary to make adjustments to chassis setup with these controls in the heat of battle. If you must spend any time at all finding the control, the distraction from what is happening on the track may cost you a position or worse. Most drivers are more accustomed to keeping the left hand on the wheel since it is usually the right hand that does the shifting. Accordingly, it seems more natural for the right hand to make these adjustments. However, many Formula Car cockpits are so tight that it is impossible to put all of the necessary controls on the right, so some concessions must be made in these cars.

It should go without saying, that the instruments should be conveniently placed so as to take as little attention as possible to read. Many club cars, however, have gauges placed far off to one side of the wheel, which requires taking the eyes far away from the road. Gauges should be placed directly in front of the driver and seen just above or through the wheel. Clock the gauges so that the normal needle position is the same for each. Then only a glance is required to ensure everything is normal. The only gauges really necessary for the vast majority of cars are tachometer, water temperature, and oil pressure. If you are concerned with oil temp, transmission, or differential temp, or the like, use temperature-sensitive stickers applied to the components in question. These can be read in the paddock after the session to determine maximum temperature reached on the track. They are also lighter and less complicated, and take no attention away from the highest priority on the track, which is driving.

The ignition switch should be easy to find, as well. Sometime during your driving career, you will encounter a sticking throttle, and it will not become evident until you lift off the throttle and begin to brake into a turn. Finding the switch quickly can make your day in a situation like this. After flicking the switch off approaching a turn, it can be turned on again at the exit while still in gear with

This is a well-laid-out instrument panel. The driver sees the instruments through the wheel, but they should be clocked so that the normal needle positions of the water temp and oil pressure gauges are at three o'clock and nine o'clock, respectively. If the needles are not horizontal, the driver knows at a glance that something is wrong. The tach should be rotated counterclockwise so that the operating range is visible.

a burst of power. This is a little hard on the drivetrain, but I know one driver who once finished a race this way.

Mirror position is also important. Mirrors should be placed so that the driver has a clear view of the areas to each side of the car and extending toward the center of the car, but the field of vision of each should not necessarily converge with the other at the rear. If a car is directly behind you or slightly off to one side, it should be visible in part in one mirror or the other. When the mirrors are positioned incorrectly, either the field of vision is incorrect or they will be too difficult to see, resulting in infrequent looks behind.

In an SCCA driving school you will most likely be beat over the head with the admonition, "Watch your mirrors." While you should always know when another car is approaching from behind, following closely or trying to pass, too much attention can be given to the mirrors, which increases the likelihood of a mistake. A very serious accident occurred at Memphis a few years ago because a driver was watching too closely as a much faster car approached from behind and neglected his own driving. This extreme is just as dangerous as not seeing a car.

A few years ago, I was working with the driver of a Swift DB-6 who was having a shifting problem. He was a good driver, and normally made very few mistakes. On this occasion, though, he had missed third gear three times in one session. We determined that the steering wheel was adjusted too close to the shifter, which made shifting difficult. On a DB-6, the wheel has three vertical positions one-half inch apart. His wheel was in the bottom position, and he thought we should move it to the top. Thinking that might be too drastic a change, however, I was able to convince him to use the center position. After the next session, he said the problem was completely solved. The moral of this story is that a very small change in control positions sometimes has a large effect. When arranging the wheel, pedals, instruments, mirrors, and other controls in your car, pay attention to the fine-tuning.

Driving a race car can be a pretty violent activity, but must still be done with precision. Being comfortable means being properly supported and having the controls properly positioned, so that your movements are quick and easy, so that you can find the controls with no difficulty, and so you can use the right muscles for the job. Taking the time to be properly fitted to the car will greatly assist in achieving these objectives.

Support Equipment

Unfortunately, the financial investment a racer has in his car is only a part of what is required to actually go racing. A number of specialized tools and pieces of equipment are needed both in the garage and at the track. In addition, a truck and trailer are needed to get it all there. We'll get to the transport equipment in a moment, but for now, let's consider the support equipment.

For working on the car, you will need the normal assortment of hand tools, which you probably already have in your garage if you are mechanically inclined. (You should be if you are to go racing.) These tools include wrenches, ratchets, sockets, screwdrivers, pliers, hammers, and the like. In addition to these, you should have an air- or electric-powered drill, a pop-rivet gun, tin snips, various types of clamps, a soldering iron or gun, a torque wrench, safety wire pliers, alignment equipment, and, possibly, an impact wrench, just to name a few. Other pieces of equipment that you will need include fuel jugs and funnels, a floor jack (or quick jack if you are running a Formula Car or Sports Racer), stands, drain pans, an engine stand, a creeper, a cherry picker (or engine lift), a vice, a bench grinder, an air compressor, and an auxiliary battery and cart, etc.

Actually, the only limit to the specialized support equipment you will use is your budget. Some club racers who are quite serious about their hobby have put up buildings in their backyards dedicated to their race cars. Many times this is to please racing wives, so that the garage is again available for the family's street cars. Others rent shop buildings and equip them with work benches, parts washer, hydraulic press, drill press, air compressor, sheet metal brake, tubing notcher, oxy-acetylene torch, and TIG welder. Some even include full machine shops with lathe and vertical milling machine. By the time they reach this level, the racers have either joined forces with racing friends to cut costs, or have gone into business to service other racers. Obviously hundreds of thousands of dollars can be spent depending on the level of commitment and finances (or financial backing) a racer has.

Although many of these items will remain on a racer's "wish list" for a long time, some of them are necessary to prepare a race car. Certain tools and equipment are also required for use at the track for maintenance, setup, and repair. The type of car you run will dictate your exact needs. Some of this equipment should be purchased as soon as you have a car. As you gain experience with a particular car, the need for other tools will become apparent. The initial investment in these tools and equipment will be high, but will diminish rather quickly. The need or desire for additional tools will be ever-present, however.

Getting the Car to the Track

Metcalf's First Law: The stuff always expands to fill the space available.

Just as the rest of the support equipment varies a great deal with requirements and budget, the tools used to get the car and that equipment to the track vary considerably in both size and price. If your racing does not include the need to entertain sponsors and their clients (although it should), transport equipment can be simple. An open utility trailer and a pickup are the norm for most club racers. Some years ago, I worked with a driver whose tow rig consisted of a relatively nice four-year-old pickup and a utility trailer, which must have covered 100,000 miles in the previous 15 years. It was rusty, had almost bald tires and the pivots for the springs were completely worn out. We even stopped at a roadside park on the way to a race to repack the wheel bearings. The driver spent all of his time and money on the race car and little, if any, on the trailer. We won the race, but most racers prefer slightly more attractive and reliable transport equipment than this.

For trailers, the next step up is usually an open trailer built specifically for race cars. These are popular with oval track and drag racers, and feature a large storage box at the front, higher than the car, with a tire rack on top. Although not too aerodynamic, they do provide storage space, which may be important if your tow vehicle is a van used primarily for moving people. Enclosed trailers are more expensive, but provide lockable storage for both the car and the support equipment. Small enclosed trailers of about 18 to 20 feet are suitable for small Production, Formula, Sports Racing, and GT cars, but GT-1, Late Model, Shelby Can-Am cars, and other larger cars would barely fit in a trailer of this size. Even with a smaller car, you may find space to be at a premium, but these trailers can be pulled with mid-sized pickups or vans. You will find trailers in the 24 to 30 foot range quite a load for a mid-sized truck. For trailers 32 feet and up, a fifth wheel or gooseneck hitch and a one-ton truck with a big engine are required. A gooseneck tongue on a trailer uses a ball mounted on the proper framework in the bed of a pickup. A fifth wheel, however, uses the same type of inclined plate used on big trucks.

Well-funded pro teams are partial to 18-wheelers. Not only does the trailer provide a great deal of space for cars, but room is generally available for the normal tools and equipment, plus lockers for driver's gear, crew uniforms, and rain gear. These have built-in generators and welding machines, and some have small machine shops or driver's lounge areas. Of all the amenities that the big rigs utilize, my personal favorite is the lift gate. Loading and unloading cars are two of the more troublesome tasks that must be performed at a race, and a lift gate makes it so much easier that one is quickly spoiled by it. Loading a car is almost pleasurable when it only requires rolling it onto the lift gate, raising it to the proper level with an electric or hydraulic hoist, and rolling it inside the trailer. Unfortunately, lift gates are expensive (around $10,000) and require a really big trailer with the structure to handle the loads.

With any size enclosed trailer above approximately 20 feet, use a fifth wheel or gooseneck hitch if you have the option. Loaded, big trailers have both a great deal of weight and large aerodynamic forces, which tend to push the tow vehicle around a lot when using a "tag" tongue design. These rigs can be nerve-wracking and dangerous to drive. I traveled with a team using a 24-foot tag-trailer behind a big, extra-length Ford van. Although the truck had plenty of power to pull it, it was an extremely nervous rig. Three hours behind the wheel was all any of us could stand because the driver was making constant small corrections to keep the trailer behind the van. On two occasions, when passing 18-wheelers, the aerodynamic forces between the vehicles caused such a tail-wagging of the trailer and the van that the truck drivers backed off and moved over to avoid being hit. More rubber on the road might have reduced this problem, but fifth-wheel rigs work much better because they introduce those forces within the wheelbase of the truck, rather than four feet behind the rear wheels, where they are multiplied.

More Support Equipment

When you first obtain an enclosed trailer, you will immediately find that you need more equipment. If you have been using an open trailer, you have been getting along just fine with only the basics. When you get a bigger trailer, though, the stuff always expands to fill the space available.

When that space is available, you will find a generator to be essential. Unless you also have room for a welding machine, it need not be a large generator. 3500 watt will do nicely for most teams. With the generator, the car's battery and the auxiliary starting battery can be kept charged. The generator will also power small electric tools, fans for hot, sultry days, and a small air compressor. Lights can then be used for working on the car after the wives and girlfriends have gone back to the hotel. A small microwave oven in the trailer is a welcome addition. Precooked meals can be heated and served to the crew and guests, and will be appreciated by all. Track food is generally expensive and not very nutritious.

Another worthwhile addition is an ATV or some other type of pit vehicle. Pro teams use custom-built electric vehicles, sometimes based on golf carts, which use a small pickup-type bed suitable for transporting tool boxes, fuel jugs, wheels and tires, and anything else that doesn't need to be hand-carried through the paddock. Since these custom-built vehicles are rather expensive, many racers use three- or four-wheeled ATVs with a small trailer for the job. The ATV can then be used at home between races for its intended purpose as a recreational vehicle. The disadvantages of three-wheelers as opposed to their four-wheeled cousins are now widely known. For racing use, with a load on the rear, they suffer from terminal understeer, and require the driver to lean forward to be able to turn.

An enclosed trailer not only allows the car and equipment to be locked up at night, but also provides a mobile workshop for servicing the car at the track. This will incorporate a workbench inside for doing gear changes or any other activity better done out of the blowing dust or leaves. A vice can be installed on the bench, and some trailers even contain a small bench grinder. Many racers attach an awning directly to the passenger side of the trailer, extending out far enough to provide shade while working. An awning with sides that zip or snap in place provides shelter on rainy days. If a generator is used, it is usually placed on the side of the trailer opposite the awning to shield the crew from the noise. The equipment you take to the track should include a roll of indoor-outdoor carpet for those occasions when your work area in the paddock is not paved. Finding that little specialized screw that has been dropped in the dirt is always frustrating.

One or more nitrogen bottles and regulators are beneficial at the track, too. The most compelling use of nitrogen is to reduce pressure buildup in hot racing tires. The air from a compressor contains a large amount of water vapor. This is even more of a problem if the compressor has not been maintained properly or if it happens to be raining. In a racing tire at working temperature, the water vapor expands more than the air, which carries it, resulting in a higher hot pressure than that set in the tire cold. Although nitrogen does still expand somewhat contributing to higher hot pressures, the effect is much less due to the lack of water vapor. Nitrogen can also be used to power air tools, reducing the need for a compressor. If you decide to use nitrogen, the bottles can be rented from welding supply stores for only a few dollars a month. The gas itself is also quite cheap. At the track, the extremely heavy bottles can be used to anchor the poles supporting the awning so that it does not blow away in a breeze.

At each event, you should have some spot in your pit dedicated to a battery-powered clock and a schedule of events. On the schedule, underline or highlight the times that you are to be on the grid so that each of your crew members will know exactly when the car must be ready. It is embarrassing and frustrating to be late for a session, especially if it is only because the crew did not know the time or schedule.

Other items that should be taken to an event are ice chests for food and beverage storage and lawn or director's chairs. Even at a club race, many more people seem to be present in your pit area than those who work on the car. This may include sponsors, their clients, prospective sponsors, friends, relatives, or other guests. If you do not feel obligated to feed them all, soft drinks would still be welcomed, and chairs will keep them where you want them, which is away from the work area. When proper seating is not available, they always seem to gravitate to the steps of the trailer, which must be kept clear as a thoroughfare into the trailer for the crew. If your racing includes sponsors, a separate tent or awning should be set up where you can feed and entertain them. At many club racing and entry level pro events, this area is adjacent to the workplace. At NASCAR and Indy Car events, special areas are dedicated to sponsor and VIP entertainment, located away from the noise and activity of the paddock.

Car Prep

Nothing is more frustrating to a driver than being in the heat of battle on the track, perhaps even vying for a win, and having some part of the car come loose, forcing his retirement. Yet mechanical failures are responsible for more early retirements from races than any other cause. Not only will DNFs (did not finishes) ruin your racing results, for a novice they are disastrous because license requirements specify race *finishes*—not just starts. It is important to learn the proper way to put a car together so that it will not fall apart.

The preparation of a racing car is a very specialized area of expertise. Proper car prep includes some knowledge of a variety of subjects including, among others, auto mechanics, air frame maintenance, hydraulics, sheet-metal work, automotive electronics, and composites. It takes a generous supply of creativity and common sense, too. To properly cover all of these subjects would require much more space than we have available here. Volumes could be, and have been, written on each of these subjects. For an excellent summary of some of them, read Carroll Smith's *Prepare to Win*. Here, we will only cover some of the highlights that are specific to race cars.

Regardless of which part of a car is being prepared, remember that if anything can go wrong, sooner or later it will. Prepare each part of the car so that it cannot possibly fail. Unforeseen failures may still occur occasionally, but with this attitude, your finishing results will improve dramatically.

Fasteners

All of the various pieces that make up a race car are held together by only four means: threaded fasteners, rivets, welding, and bonding. The most common of these, of course, are threaded fasteners—nuts and bolts. A fastener can be used in either a stressed or nonstressed application.

Examples of stressed fasteners include suspension fittings, brake caliper and master cylinder mounts, engine and transmission mounts, clutch and flywheel bolts, wing mounts, and similar applications that are not only important, but subjected to severe loads as well. Nonstressed fasteners hold the radiator and battery in place. Even though a failure of one of these nonstressed fasteners will not have catastrophic consequences, it may still cost you a race.

Hardware store bolts are certainly not adequate for stressed applications, and are not advised for nonstressed applications, either. Even bolts bought from industrial fastener suppliers are suspect these days since many found to be inferior are coming in from overseas. The SAE grading system is still sound, but many counterfeit fasteners have arrived on our shores using the SAE grade markings, but not meeting SAE specifications. To be sure about the bolts you use to put your car together, use only AN (Air Force-Navy), MS (Military Specification), or NAS (National Aerospace Standard) fasteners. If purchased from first-run aircraft suppliers, such bolts are very expensive. However, fortunately for us, the U.S. government has had these made by the millions since the '50s, and large numbers of new, unused bolts, nuts, and washers have found their way to surplus houses. There, they are dirt cheap.

The bolts to use are the AN-3 through AN-10 series, where the number designates the diameter of the bolt in sixteenths of an inch. The grip length (unthreaded shank) of these bolts varies in 1/8 inch increments from 1/16 inch to six inches, which should be enough for almost any application. These bolts are only available with fine threads.

In addition to being used in stressed and nonstressed applications, bolts can be used in tension, single shear, or double shear. Any bolt, when tightened, is loaded in tension, but shear forces are sometimes introduced, too. The head bolts of an engine are tensioned fasteners. Single-shear fasteners include clutch and flywheel bolts, U-joint and CV joint bolts, and wheel studs. The shank of the bolt is designed to be used to withstand these loads. The threads are provided to load the bolt in tension and keep it tight. When threads rather than the shank are loaded in either tension or shear, they provide stress concentration points. This reduces the load-carrying capability of the bolt and may cause a failure. Make sure that every bolt you install loads only the shank in tension or shear, not the threads.

Course threads should not be used on a race car—except in pre-threaded blind holes such as motor mount holes in engine blocks. Some people believe that course threads should be used in aluminum and magnesium, and it is true that a course thread is stronger, but it also has less holding power. The aerospace industry uses fine-thread stainless steel thread inserts for *all* threaded holes in light alloys, and there should be a lesson for us here. Course thread fasteners have no place on a race car, except where they cannot be avoided. When they must be used, AN-73A to AN-81A bolts are the proper choice.

All of these AN bolts, have a small radius under the head where it meets the shank. This radius reduces the stress in this area of a bolt in tension, and therefore increases its tension capacity. To accommodate this radius a corresponding washer, AN-960, must be used.

Bolts of even higher strength than the AN series are available. These are MS20004 and NAS624. They have an increased radius under the head and require a chamfered washer, an MS20002C.

Some racers frequently use Kay nuts, also known as Jet nuts. These are small, giving good clearance around the head; all steel, so that they can be used in high temperature applications; very strong; and self-locking. Their disadvantages are that they tend to be hard on the male threads. They are expensive, about $.50 each for a 3/8, a price that jumps to about $3.50 each for a 7/16. Other racers mainly use them on header studs, brake rotors, and CV joints, where their size is a big advantage. For general-purpose use, elastic stop nuts may be a better choice. Again, not the hardware store variety, but the higher-quality AN-365 nuts. Regardless of the nut used, it should have an AN-960 washer under it. When tight, at least two threads should protrude past the nut.

Any fastener on a race car absolutely *must* use some sort of locking mechanism. If it does not, you are courting suicide. The vibration forces and frequencies demand this. Lock washers are not adequate. The quality control used in their manufacture leaves something to be desired, just as with some bolts, as many are too hard and break leaving a loose bolt. Others are too soft and lose their tension, allowing vibration to loosen the fastener. Elastic stop nuts are adequate and usually do a fine job. They are designed to be used only once because they are a bit softer than the bolt, though. When one of these nuts is removed, throw it away. Do not reuse it, and don't throw it into a box of junk nuts and bolts because someone else might dig it out and reuse it at a later date. Get rid of it. New nuts are cheap insurance.

I once had a car that had a used elastic stop nut on the pivot bolt of the brake pedal. During a test session, it came off, leaving the brake pedal unconnected to the chassis. Murphy (of Murphy's Law fame) prevailed, and it occurred at the end of the longest straight. The car rolled numerous times, completely destroying it. The driver was lucky to escape with only minor injuries. I now use a castellated nut and a cotter pin in this extremely important application. Many other applications call for safety wire. Loctite can be used to prevent vibration from loosening fasteners in blind holes, but the directions on the bottle must be followed religiously to get acceptable results.

The best way to lock bolts in blind holes is with safety wire. Obviously this requires that the bolts have drilled heads. This is a pain if you must drill them yourself, but some AN bolts come with drilled heads. Buy them when you find them; the hole does not weaken the head and is very handy when you need it. Safety wiring is not difficult, but must be done properly to be effective. Although it can be done with regular pliers, safety wire pliers will allow you to do a better-looking job much faster. Stainless steel safety wire comes in several diameters, but .020 and .032 inch are the two you will use most often. Always arrange the twisted wire so that a loosening bolt will put the wire in tension to prevent the bolt from turning.

One final note about threaded fasteners. Many years ago a former teammate from my Karting days modified the front end of his Kart, which required shortening the tie rods. On this particular Kart they were male rods fitted with female rod ends. When he shortened them, he cut off the threaded portion and cut new threads using a die. He did not realize that the original threads had been rolled to the proper diameter from a slightly undersized rod. The new threads, then, were not full diameter. This caused no problem for about two years, but the rod finally pulled out of the rod end leaving him with steering on only one wheel. Murphy prevailed again, and the resulting understeer occurred in a high-speed downhill corner. When he hit the outside guardrail at a high rate of speed, he was killed. All methods of joining race car parts are important, but we use more threaded fasteners than any other type. When using them, pay attention and observe the rules. It may lengthen your racing career and your life as well.

Electrics and Hydraulics

Failures due to electrical problems usually occur as a result of one of three causes, the most common being vibration. The area of the car that sees the most vibration is adjacent to the engine. The vibrations here are due to the reciprocating parts of the engine changing direction, and are low in amplitude, but very high in frequency. Any electrical components located in this area will be subjected to oscillatory forces that will fatigue mounts, electrical connections, and internal electrical components. For this reason, it is not advisable to mount a component such as an ignition coil or fuel pump directly on the engine. By using solid engine and transmission mounts, which are preferable, this vibration can be transferred into other parts of the car as well. All electrical components should use some sort of isolation mounts.

The brackets to which all the other components mount should be designed so that they will not fatigue over time. Brackets that fail will end a race just as surely as a component that fails. Keep in mind that the bracket mounting a component must not only hold the weight of the component, but must also be sturdy enough to withstand the G forces involved in cornering, braking and acceleration. It must also withstand the forces experienced when wheels are bounced off the track in bumpy sections, when hitting FIA curbs or when dropping wheels off the track. Cornering and braking forces, even in club cars, can sometimes top 2 Gs, which means the bracket must be able to support twice its normal load. The forces on the suspension can sometimes momentarily exceed 10 Gs, so components mounted in this area are subjected to intense punishment. As an example, some older Formula Fords have had batteries mounted above the gearbox. In this location, a battery receives vibration from the engine and the suspension, and its mountings have very short life spans. A better battery location on a Formula car is somewhere in the driver's compartment, where these forces have been somewhat dampened by the frame and other attachments of the car.

The second most common cause of failures due to electrical problems is connections. The usual method of attaching a terminal to a wire is by using the popular crimp connectors found at auto parts and hardware stores. Although these do work many times, the failure rate is unacceptably high. There seem to be some quality control problems with these connectors, but, by far, the most common problem is operator error. Many people are guilty of using the wrong size connector for the wire, or of undercrimping or overcrimping the connection. Undercrimping will obviously leave the wire loose in the connector, and overcrimping puts a stress riser in the connector, which vibration will eventually attack. Aircraft mechanics use crimped connectors, but use Mil Spec parts and crimping tools that are designed to avoid both of these situations. If you are going to use crimped connectors, such parts and tools are worthwhile investments.

I prefer to solder terminals to wires. Even this can be done wrong. A soldered joint can be overheated, which will weaken the wire. If it is underheated, a cold joint will occur, which is high in electrical resistance. Practice and awareness are required to solder a joint correctly, and, once finished, a soldered joint should be protected with a length of shrink tubing to insulate it and give it some extra strength. It is very frustrating to find that you lost a race because a connection on a coil or switch failed. Regardless of the attachment method, only ring terminals should be used, not the forked type. Spade terminals are sometimes required, but should be secured with .020 safety wire through a small hole in the mating parts.

The third most common cause of failures due to electrical problems is crimped or chafed wires. If the car is bottomed on the track or trailer and wires are routed along the bottom of the frame or unibody, they will be damaged. If they have been run in locations where parts are moving, such as the suspension area, it is possible that they will be crimped. This is also common in the engine area, where wires are crimped when changing engines. Another trouble-prone area is where wires are routed adjacent to AN hoses. The stainless steel braid works like a file on everything it touches, and will quickly abrade away insulation on wires. Finding places to route wires, where they will be

Wiring causes more DNFs than any other mechanical cause. The possibility of failure was high enough when Al Holbert was running this Porsche 962, but with computers controlling engine functions today, the possibility of an electrical failure is even greater, requiring attention to wiring detail.

out of harm's way in all of these situations, is sometimes difficult, but worthwhile. Remember, anything that can go wrong sooner or later will. Time spent in preventing problems in this area will make your racing more productive and enjoyable.

As a final note on electrical wiring, following is a word about fuses. Many amateur cars are wired using fuses, but if the wiring is done properly, fuses are not required. A fuse is only necessary if something goes wrong. As we have seen, the problems that a fuse would normally remedy result from a wire or terminal shorting due to improper wiring procedures. Occasionally, however, component failures occur, such as a fuel pump shorting internally and drawing excessive current, which would burn a wire if a fuse were not present. This is an extremely rare circumstance, though, and the race would not be finished regardless of the cause, whether a burned wire or blown fuse. (Actually, the engine would quit first in this instance due to a lack of fuel.) Automotive fuses themselves have a high potential for failure due to the design of their connections. Very few pro cars use fuses, and their finishing rate is significantly higher than amateur cars. We can safely assume that the use of fuses does nothing to further that cause. Having said that, the plug-in plastic European fuses that are now in use are much more reliable than the American glass variety, and I do not object very strongly to a car constructor wiring the car using them.

In spite of the fact that a race car's hydraulics are very simple, they still cause more than their share of problems. Except on Showroom Stock and some GT-1 or Trans Am cars, the only hydraulics involved are the clutch and brakes. The bigger cars sometimes use power steering as well. Hydraulic lines deserve the same protection as you would give to wiring. Don't allow anything to crimp or abrade them. Route the lines out of harm's way, so that they will not be crimped during engine changes or off-track excursions.

The most common problem occurring with brakes and clutch hydraulics involves not changing the fluid at frequent intervals, and not rebuilding the masters and slaves often enough. Brake fluid is hydroscopic and will absorb water from the atmosphere. The water trapped in the fluid lowers the boiling point and promotes corrosion. During dry conditions, it is acceptable on most cars to bleed out the old fluid prior to each race. When raining, this may need to be done two or three times in the same weekend. On high-powered or heavy cars, where a great deal of brake heat is generated, more frequent fluid replacement is necessary. Failing to replace fluid frequently enough results in lowered ability to withstand heat and rust and/or corrosion in master cylinders, calipers, and slave cylinders. If the car is parked at the end of the season and the hydraulics are not touched again until spring, rebuilding of the cylinders will be mandatory. On low-powered cars, rebuilding these parts once a season is usually sufficient. On faster cars that produce a lot of heat, it sometimes takes once a race to keep everything working properly.

Light, low-powered cars can easily use Castrol LMA or other readily available brake fluid. Heavier or faster cars must use a fluid with a higher dry boiling point, such as AP 550 or 600 or Wilwood 570. As an example of what can happen if the wrong brake fluid is used, I was once at a race at Elkhart Lake, a track that is particularly hard on brakes, where the driver boiled the brake fluid. His explanation for coming in during a session was, "No brakes." We connected the car to the ATV to pull it into the paddock, and, sure enough, *no brakes!* The pedal bounced off the floor.

The car would not even stop the ATV. When fluid boils, it turns into a gas which compresses and does nothing to slow the car. On low-powered cars, the heat generated is not usually high enough to cause this problem. On faster cars, though, it is an ever-present challenge. Silicone fluid should never be used in a competition application as it has some major disadvantages. According to Mac Tilton of Tilton Engineering, "Silicone brake fluid which meets DOT 5 specification is unsuitable for high-temperature applications because of its compressibility. Silicone brake fluid is not being used in any series where brakes are subjected to high temperatures, such as with carbon/carbon brakes."

Suspension

Maintenance of suspension is relatively straightforward, but still gets less attention than it deserves. Almost all maintenance of these assemblies involves bearings. Tapered roller bearings, sometimes called Timken bearings, are common on many cars as wheel bearings. They should have a very carefully sized spacer between the inner and outer bearings in order to maintain proper adjustment. When this spacer is the proper length and the nut is pulled up against the outer bearing, both bearings and the spacer are placed in compression on the spindle with the hub and races floating on it with a minimum of play. Without the spacer, the bearing/hub assembly does not have as much load-carrying capacity, and the bearings must be overtightened, increasing friction between the bearings and their races.

Some cast uprights use two angular contact ball bearings pressed into the casting. This system, too, requires a spacer to preload only the inner races, so that the balls and outer races are not unduly loaded. These spacers should be resized whenever bearings or races are changed. Many newer cars use only one wheel bearing in the upright, a double-row angular contact bearing. The only maintenance these bearings require is making sure they are properly sealed and adequately lubricated. Brake heat transferred into the uprights and bearings sometimes damages the seals. This is evident by the radial tracks the grease leaves on the side of the bearing. Bearings with worn, pitted, or flattened balls, rollers or races are found by examining the individual parts or by rolling a clean, grease-free bearing in your hands. Any rough spots are cause for replacement.

Suspension bearings on production-based cars are many times the same ones used on the street versions of the car and should receive the same maintenance. These are usually ball joints, tie rod ends, and rubber pivot bushings. More serious suspensions will use spherical rod ends, which are sometimes called "Heim joints." These bearings are pretty trouble-free, but should be replaced when any play can be detected in them. This will first be evident axially, in line with the bolt, and will only be noticeable radially when wear has progressed to massive proportions. Cheap rod ends have axial play when they are new and should not be used. A typical front corner of a Formula car has at least eight rod ends, and a little slop in each of them can grow into a major problem. Careful suspension alignment can be negated by sloppy rod end bearings. The best ones have Teflon-coated races and can barely be moved by hand.

Rod end bearings are designed to have freedom of movement in two planes, at an angle to the axis and in rotation. Failures occur frequently because they are installed with insufficient travel available. This will damage the race or bend the shank. To allow enough movement, install them with tapered spacers on each side. Any bearing used in single shear should have a tapered retainer washer at the outside to prevent it from coming loose if the ball breaks or pulls out of the housing.

Engine

The engines found in race cars come in only three varieties: racing engines, production based engines modified for competition, and stock engines. Formula Fords and 2000s use stock engines. Modified production engines are found in Production and GT cars, and racing engines are used in Indy Cars, Formula One, F/3000, etc. The type of engine used in a car will have a large effect on the maintenance it requires. Pro teams use fresh Ilmors or Cosworths at each race. Sometimes qualifying and/or practice engines are also used in the same weekend. The mechanics earn their keep. For almost all other types of engines, several weekends can be expected from each engine between rebuilds. The engines used in the spec engine classes, i.e., Spec Racer, Formula Ford, Formula Vee, etc., are largely built and freshened by professional engine builders at four- to seven-race intervals. Production and GT engines are usually built and freshened by the driver in his garage at similar intervals.

To properly determine when an engine requires a rebuild, leak down tests should be performed after each event. A leak down test involves moving each piston to top dead center with the valves closed, applying compressed air to the cylinder, and determining how much leaks out. Leak down figures of two to three percent are good for a new engine; five to seven percent is average for most engines; and a rebuild should be considered at around nine to ten percent.

Leak down tests can determine not only how much leakage a cylinder has, but the cause of that leakage, as well. If a high percentage of leakage is due to improper ring seal, it can be heard as air escaping into the crankcase; intake valve leakage can be heard at the carbs or injectors; and exhaust valve leakage is detected at the header.

An internal combustion engine is an air pump. The more fuel-saturated air it flows, assuming the fuel is completely combined with the oxygen in the air through combustion, the more power it makes. In the early part of an engine's productive life, horsepower loss is almost always due to leakage past the valves. In the '60s, Mark Donohue worked extensively with Traco Engineering on Chevrolet V-8s. His conclusion was that lapping the valves after the engine was broken in increased power by about 20 horses! This amounts to bringing the leak down figures down from the 5 to 7% of most engines to the 2 to 3% of a new engine. Smaller engines, of course, would show proportionately lower gains. Many racers have discovered this fact, and it is common at many entry-level pro races (F/2000, etc.) to see mechanics pulling the head and lapping valves the night before the race.

On high-revving engines, the valve train requires constant attention. Not only will the valve lash need frequent attention, but the valve train parts themselves require periodic replacement. Teams running Chevys, for example, must replace rocker arms, pushrods, and roller lifters at intervals ranging from 600 to 1000 miles. Neglecting replacement of these items is courting disaster. Maintenance is also required on ignition systems (usually the mechanical parts) and on carbs. Each type of engine is a bit different and requires its own program of maintenance. Engines are sometimes like kids. Learn the personality of your engine so that you can give it what it wants before it throws a tantrum.

Gearbox and Final Drive

Most racing gearboxes are pretty straightforward. Upon examining one, it is easy to understand how the power is transmitted from one shaft to the other, and how the shifting is accomplished. Care must be exercised, however, to ensure proper operation of the bearings, the preload on the bearings, and the proper mating of the gears. Study diagrams of the workings of your gearbox and study its parts to become familiar with the theory behind its operation, as well as how the individual parts work together. Once you have this understanding, proper maintenance of the gearbox will become almost second nature.

Low-powered cars can use pretty lightweight transmission oils. Formula Fords typically use a light synthetic such as that manufactured by Red Line. In the early '70s, we used 85-90W Castrol, Valvoline, etc., in these cars. In the mid-'70s, someone discovered that 30 weight motor oil worked better. Even later, we discovered that light synthetics allowed the box to cool significantly faster after a session, and this implied that less heat was generated in the first place. Less heat generated and thrown off into the airstream means more power to the wheels.

In higher-powered cars, both the weight and the temperature of the gear oil have a significant impact on gear longevity. This is particularly true of ring and pinion gear sets in which sliding motion is predominant. In the later stages of the second incarnation of the Can-Am, Chevys were producing about 580 BHP. Downforce was significant, and the DG-300 Hewlands were only marginally up to the task. The individual ratios were not a concern. The ring & pinion, however, were changed after each race. Heavier oils and/or better gearbox oil coolers were needed, but were not available or not foreseen. The real fix was a larger ring & pinion, which was soon available with the DGB Hewland box.

Preparation is the key to finishing races. As the saying goes, "You can't win unless you finish." Many novice racers do not understand the importance of proper race car prep, or they are not aware of the "nuts and bolts" rules of car prep. This lack of knowledge causes many early retirements from club races. At the pro level, while the cars are generally more highly stressed, they also receive more care. Mechanical failures still occur, but are usually slightly more obscure, such as a valve that breaks in operation taking the rest of the engine with it, or a lug that breaks off a battery and leaves the car without electrical power. It is impossible to prepare a car so as to absolutely preclude failure, but that impossible goal is worth striving toward. Doing so will give you the best chance possible of finishing races, and that is the first step on the road to winning.

Chapter 4

SCCA 101

If you like the federal government, you'll love the SCCA.

Grassroots racing in America is done on short ovals, for the most part on Friday and Saturday nights. These tracks are either dirt or pavement, and each type of surface is popular in its own parts of the country. A reputation follows the competitors who run on these tracks regardless of the type of surface. They are regarded by most outsiders (even the fans) as somewhat uneducated, aggressive brawlers. Perhaps this has something to do with the thinking that people only watch races to see the crashes (or the fights afterwards). Although this reputation was once well-deserved, the competitors themselves have cleaned up their acts. Arguments in the pit area still sometimes ensue, but real fights are rare. Driving-related disagreements don't only happen at short tracks, though. They sometimes happen at longer ovals and road courses, too. Some may remember the punch thrown by A.J. Foyt that connected with Arie Luyendyk after a 1997 IRL race. The difference is that in pro racing, competitors must answer to a national sanctioning body, and the one most competitors are likely to encounter first is the Sports Car Club of America (SCCA).

If you race in North America, sooner rather than later, you will run with the SCCA. This organization sanctions most of the road races and some of the oval races in this country, both amateur and professional. Since SCCA is such a dominant organization, and since it has some peculiarities and hidden pitfalls, it is important to know a little about its structure. However, lest you think I am picking on SCCA, let me say that I am simply using this organization as an example. Other racing organizations have their own pitfalls, which you should get to know as soon as you begin to run with them.

Most racing organizations are similar. Jim Edwards, long-time racer, track owner, and team manager once exclaimed, "The officials in pro racing aren't any better than the ones in amateur racing, but you get paid to put up with them!" It is important to learn how to get along in the club with which you are racing. The lessons you learn with the first organization will also apply to USAC, PSR, IRL, CART, FIA, and NASCAR.

Organization

If you like the federal government, you'll love SCCA. Like the government, SCCA headquarters in Denver tries to please all of the people all of the time. However, this is a difficult if not impossible task, and many members are left perpetually disappointed. As a result, there is a duality within the club, with the administration on one side and some of the membership at large on the other—only "some" because most of SCCA's 53,000 members are not racers. Many drive Rally cars, some run in Solo competition, some own sports cars and simply enjoy belonging to the world's largest car club, some are wives and family, and some just join to receive the Club's magazine, *Sports Car.* Thus, the group that is usually at odds with Denver is comparatively small. But their investment in terms of equipment and lifestyle is large, and they become somewhat emotional when they feel that SCCA is meddling in what they consider to be their business and their lives. Particularly distressing are situations where one tries to find an advantage in the rules and spends a great deal of time and money to put the car on the track only to find SCCA changing the rules after the fact and outlawing it.

The other side of the coin is SCCA's unenviable position of being "mother hen" at the same time as it is "hungry fox." SCCA writes the rules, runs the races, enforces the rules, insures the competitors, and generally does a myriad of things with which racers do not want to be concerned. At the same time, their relationship with racers is frequently adversarial, when rules and policies are not changed as rapidly as the competitors feel is required, or when protests or disqualifications are brought against a competitor by the Club. But the Club is the only game in town, so some racers need it and despise it at the same time.

For the racer, getting along in SCCA takes some schmoozing. First, the people in power must be identified, and their particular duties and areas of control must be ascertained. The latter is important because some tend to rule with an iron fist when given authority over racers. Many of these are normally good people, but few are paid Club employees. Most are volunteers who give up their weekends to stand in the hot sun, walk a lot, and have irate racers argue with them. So, in light of their circumstances, maybe their behavior is not so surprising. Just remember, though, if you argue with the tech inspector on a regular basis, he will look for a reason to throw you out. Take him a beer when the day is over; make him your friend. The same goes for all other officials, from corner workers to the Director of Club Racing. If you plan to race for long, make friends, not enemies.

SCCA has several Divisions, each for a different part of the country, and each Division includes several Regions. Races are conducted by the Regions, and points from National races are totaled in each class for each Division. At the end of the season, the driver with the most points in a Division in his class is the Divisional Champion. However, this honor generally does not mean much. Many times a slow driver will win the Championship just because he makes all of the races, while a faster driver from the same Division will score fewer points, but go to the Runoffs™ and do well.

Although your membership is with a particular Region, you will be required to register with a particular Division. Your points will be scored in that Division, but you may also include points from two out-of-Division races. You are not required to join the Region, or run with the Division, in which you live, though. Some members join a Region they visit frequently on business, while others join the Region they feel has the best monthly publication. One driver listed his Division of Record as the one he felt had the best competition and best tracks. Investigate the possibilities and join the Region and run with the Division that seems best for your circumstances.

Each Region elects its own leaders. SCCA is a rather "cliquish" organization (as are most other groups of all kinds), so it will be to your advantage to be friendly with the group in power. Most racers, perhaps because of a combination of personality traits (go back to Chapter 1 for more information on this), seem to be uncomfortable with the administrative side of racing. They prefer to be on the track rather than behind a desk. The reality, though, is that the politics of racing dictate that a driver should be at least somewhat involved in the administrative side to improve chances of long-term success. This does not mean that you cannot win races if you never go to monthly Region meetings, but being on good terms with those who make and enforce the rules and policies will go a long way toward promoting smooth sailing off the track. It is common for racers to feel as if they are being "hassled" by tech inspectors or penalized by the Club, or are having protest decisions go against them for no apparent reason. Sometimes the reason is merely that the racer is not a member of the ruling clique.

Dealing with Officials

As a beginning driver, many of the things you must do to fulfill license requirements will require contact with officials. To get a Novice Permit, you will be required to speak with your Divisional Licensing Chairman and those in your Region or in Denver in membership. At your driver's school, you may need to speak with the Chief Steward to get him to waive your second school so you can run a Regional race. Finally, your car must pass an annual technical inspection, and you must deal with your Region's Tech Inspector for this.

Following are job descriptions of some of the officials you are likely to encounter:

Chief Steward This person is responsible for the actual running of a race event. He is the boss to whom all the other race personnel answer and he has ultimate authority at the event.

Stewards of the Meet These guys are the judge and jury in protest situations. They are not involved in the organization or running of a race except to modify the schedule or change the Supplementary Regulations when necessary. They answer to no one except those in Denver.

Race Chairman The Chairman is mainly responsible for planning all the aspects of a race, i.e., leasing the track, making sure adequate staff is available to run the race, obtaining emergency vehicles, etc.

Registrar The Registrar's job is to accept and process all race entries and ensure that all competitors at an event have current SCCA membership and the proper licenses.

Divisional Licensing Chairman This individual has authority over the issuance of all licenses within a division.

Regional Executive This is SCCA lingo for the President of a local Region.

Starter This guy shows the green flag at the start of a race and the checkered at its conclusion. Learning how he starts races may give you an edge.

Corner Workers These dedicated and hard-working people stand out in the hot sun all weekend showing you various flags, pushing your car out of the way when you stall or spin, and offering assistance in the event of an accident. If you do spin or run off the track in their area, your car belongs to them. They have the responsibility to get you back on the track safely or park your car in a safe place. Do what they tell you.

Driver Observers These individuals give written reports of any on-track incidents or accidents to the Chief Steward.

Sound Control Judge This person monitors the noise level of individual cars on the track. If your car is too loud in a practice or qualifying session, he will inform you.

Chief Timer and Scorer This person and his or her crew times every car during qualifying sessions, and keeps lap charts during a race showing each car's position at any time during the event. From time to time, they do make mistakes. When that occurs, you may plead your case with Timing and Scoring, but you must be able to back up your claim with hard data and evidence. You can be sure that they will.

Technical Inspector (Scrutineer) Tech Inspectors are responsible for inspecting a race car and/or driver's gear for compliance with safety rules prior to an event and for compliance with specifications after an event. They can require a driver to install different seat belts prior to a race and disqualify him for having too large an engine after a race.

These people have the power in their respective fields to make your racing pleasant and enjoyable, or they can be your nemesis. Their jobs within SCCA are to ensure that the racing is fair and the events run smoothly. While many of them are intelligent and open-minded, and use common sense, a few are apparently subjugated during the week by a domineering boss or spouse. These "tin gods" seem to relish using their weekend authority to make the racer's life at the track a living hell. Treat them all with respect, but identify those who require special handling.

Rules

In American motorsports, the classes that have traditionally enjoyed growth or steady, high participation, Sprint Cars, Formula Ford, and Formula Vee, have rules that have been fiddled with very little for a very long period of time. Fortunately, SCCA seems to feel that their two classes are doing just fine, thank you, and need no rule changes. In most other classes, however, change is the essence of racing. Sometimes rule changes are needed and, after overcoming a lot of inertia, SCCA obliges. For example, the Production Car rules received a major overhaul in 1992, whereby many rules were streamlined and many others that were just plain out of date were replaced. The old rules had been involved in controversies for many years, and the new rules, while not perfect, went a long way toward being easier to understand with less "interpretation" involved in enforcement.

Rules are needed in racing to provide a level playing field on which to compete. To that end, they should be clear and concise, and spell out exactly what is and is not allowed. The difficulties involved in creating rules to do this are twofold. First, it is the nature of the sport that racers are always looking for a performance advantage. Competing in completely equal and identical cars may be the ultimate contest between drivers, but if the rules allow any creativity at all in hardware, the door is opened to technical innovation. This makes it a contest not only between drivers on the track, but between designers, engineers, crew, and everyone else involved off the track. It is inevitable, therefore, that a contest ensues between the racers who are looking for advantages and those who write and enforce the rules. Racers are always looking for (and finding) technology or creative engineering that enables them to go faster—technology that the rulemakers may have never even imagined.

The other difficulty arises in categories that use production-based automobiles such as GT and Production Cars. Many times the cars eligible in a class are so diverse that it is virtually impossible to create logical rules to fit them all. Consider the problems of generating rules to fit open, closed, front-engined, mid-engined, and rear-engined, front-wheel-drive and rear-wheel-drive cars into the same rule framework. This is what the Comp Board must do in the Production Category. If a rule states, for example, that motor mounts are free, but also states that front engine plates cannot be modified, and we find that a certain car uses motor mounts integral to the front plate, a conflict exists. Gray areas such as this in the rule book lead to differing interpretations of the rules by different Tech Inspectors (and competitors), which leads to different outcomes of similar offenses on different occasions.

This difficulty could be overcome if SCCA recognized precedents in their rules enforcement. This is one major difference between Washington and Denver. Our legal system is designed so that the outcome of a case can be brought to bear on the finding of a similar case. The reasons that lawyers have many books in their libraries is so they may look up similar cases and present those decisions to the judge. In SCCA, however, the outcome of a particular rules infraction may go one way on one occasion and the opposite way on another. If rules cannot be written in such a way as to be clear and unambiguous in all situations, standards should be established to promote consistent rule interpretation. If the Stewards of the Meet were required to consider the findings of similar protests when ruling, enforcement would almost certainly be more consistent.

On some occasions, rule changes have more to do with administration than with racing. Through the late 1980s, Production Car rules allowed an asymmetrical roll cage, i.e., a high roll bar behind the driver, sloping down to door level on the passenger side. In 1992, a new requirement was added whereby an "intrusion protection" tube had to be added to the passenger side of the car. When asked why this new requirement was written in, the Technical Director of Club Racing replied that, although they could cite no injuries, they were afraid that a driver would be injured from an impact from the passenger side. When asked why this was not a problem in Formula Car cockpits, where the driver sits against both sides of the car, there was a long silence—and no answer.

Each year at the Runoffs™, a meeting is held with the Competition Board, who writes the rules, and the competitors in each category. At the Production Car rules meeting at the '92 Runoffs™, the first question asked by one of the competitors was, "Will the roll cage rules remain as they are for a few years or will we soon have to modify our cages again?" The reply from the head of the Comp Board was that they would remain as written. In '93, the rule was changed to require a full-height, full-width roll bar. The new Club Racing Technical Director, when asked about the new rule, said that although a full-width roll bar was now required, the rule was being overlooked. Rules that are overlooked are just as ambiguous and can be as unfair as those that are enforced, but not written. Without consistent, stable rulemaking and rules enforcement, long-term longevity in a class cannot be ensured. Who wants to build or buy a car that may not be legal next year? Unfortunately, that is a way of life for most club racers.

Rules are sometimes changed with a positive result, however. One rule concerning suspension was written in such a way as to preclude any suspension modification past changing an anti roll bar. In another area, it was clearly stated that any suspension modification was legal as long as the suspension type was unchanged. These conflicting rules are the type that prompt overzealous tech inspectors to key on the one rule they know best and ignore the other. Another tech inspector may do just the opposite. Obviously, this is not a healthy situation for the racer or for the Club. When this conflict was pointed out to the Comp Board, one rule was changed and the conflict disappeared.

Occasionally, rule changes in pro racing are a result of series sponsors reaching their goals (or sometimes not reaching them). Volkswagen of America pulled their dollars out of the Formula Super Vee series a few years ago and, when no new sponsor was on hand, the series folded. Shortly after many racers had installed Olds Quad Four engines in their Sports 2000 cars, Oldsmobile decided not to continue to support the series they created. The same thing happened to the Shelby

Can-Am pro series competitors. After creating both amateur and pro classes and selling only a few dozen cars, SCCA realized it was not catching on and dropped the pro series. Buyers of these relatively expensive cars are now trying to decide what to do with them now that they are worth about 25% of their original price.

If all this talk of constantly changing rules makes you feel somewhat uneasy, you have good reason to be. Racing is a rough sport on and off the track, and those who succeed are able to move quickly and roll with the punches. This is not a sport for the faint of heart. If you know in advance how the rulemaking and enforcement system works, you will know what to watch for and what to watch out for. Being able to spot these political obstacles will help you to choose your off-track lines.

In Washington, the legislative branch makes the law and the judicial branch enforces it. In Denver, the Competition Board makes the rules and the Tech Inspectors, Stewards, and Appeals Board enforce them. While this may be best for 230 million people, some in the Club feel that a single legislative/enforcement branch might be more appropriate for keeping competition fair. When one group writes rules and has no say in their enforcement, those enforcing the rules must try to determine the intent of those who wrote them. More often, however, those enforcing the rules "interpret" them to suit their own ideas.

An even more radical idea is for the competitors themselves to have a say in the rules they race under. Presently, there is within each competition category an "ad hoc" committee made up of racers who make recommendations to the Comp Board. This sounds like a perfect arrangement, but the committee is appointed by the Board, not elected by the racers, and seems to more resemble a figurehead than a practical organization. An arrangement in which the racers themselves have a voice in the rulemaking process seems to work quite well in NASCAR and Indy Cars, but Denver would probably strongly resist this. The SCCA is, though, a club of the racers, by the racers, and for the racers, so it will be interesting to see if this growing sentiment within the ranks exerts enough influence to make a difference.

Licensing

Before you can race with SCCA you must have a license. If you have no previous racing experience, one of the best ways to get started is through SCCA's club racing program. To do this, the General Competition Regulations, or GCR, say that you will need to complete two SCCA race driver's schools before you can compete in your first Regional race. An accredited professional school can be substituted for one of these. (See Appendix A for a list of accredited schools.) In reality, only one SCCA school is required, regardless of what you have done before. After you complete your first school, request of the Chief Steward that your second school requirement be waived. Provided you do not do this in an antagonistic way, and that you have shown yourself to be a competent driver, he will probably grant your request. Since most drivers's schools are conducted as a part of a School/Regional weekend, this will let you complete the school on Saturday and compete in the Regional race on Sunday.

Your school and first two Regional races will be run under authority of your Novice Permit. Upon completion of these races, you can exchange the Novice Permit for a Regional Competition License. This is not absolutely necessary, however. Four more Regional races must be run to obtain a National Competition License, but they may be under a Regional License or your Novice Permit.

For those having little or no previous on-track experience, SCCA's licensing program is very well thought out. The schools provide the officials an opportunity to watch a driver to make sure he is safe and competent. And the schools provide the driver quite a lot of track time in one day—sometimes more than three hours—which is good for getting used to a competition driving environment. The Regional races also provide a considerable amount of track time, during which the driver can hone his skills. Thus, by the time he has qualified for a National License, he is probably ready for it.

Pro licenses with some major organizations are comparatively easy to get with only a modicum of amateur experience. After obtaining your National License, it is a simple matter to fill out the appropriate forms and send in the required fee to get an SCCA Pro License. Sometimes other requirements must be met, such as running three additional National races, but that is not always the case. One driver who had only limited Karting experience ran one school and one Regional race, was granted a National License, and then applied for and received a Pro License. Being friendly with the proper officials helps. Before you try something like this, however, be sure you are ready.

Organizations such as PSR, IRL, and NASCAR usually look at a driver's résumé and credentials (licenses) to determine if he should be given a track test. If they believe his experience shows he could be a safe and reliable driver in their kind of cars, they will give him the opportunity to test. I know one driver who had only Karting experience and applied to IRL for a license. One of the officials told me, "It just isn't going to happen."

When running Formula Atlantic, Price Cobb read in a magazine that he had been granted an FIA Super License, which would qualify him to run anything, even Formula One. Apparently, his name had come up in an FIA meeting and he was accepted without even knowing he was being considered. Pro licensing procedures vary among the organizations; with some it is easy to qualify, with others you must really show your stuff. Contact the licensing administrator of the organization with whom you want to run before it is time to apply, so you will know what they require.

SCCA Schools

Before entering your first school, you must be an SCCA member and complete the requirements for a Novice Permit. Call SCCA in Denver or see the Registrar at a race for the necessary forms, which include a membership application, a medical form, and an application for the permit. Of these, only the medical form could cause a problem. It is much too common for the beginning driver to wait until the last possible minute before scheduling an appointment with a doctor for a physical and have to hurry to a fax machine to get the completed form to Denver. Get all of the required paperwork done ahead of time and send it in early. You should receive your Novice Permit from SCCA and have it in hand when you send in your school entry to the Registrar.

Do not expect to learn much about competition driving in an SCCA school. It's not that the instructors don't try to teach or are not competent—they do and they are. SCCA schools are just not set up with driver development as the primary concern. Their main function is to let the instructors and Stewards view you on the track. Anything you learn should be considered a bonus. For that reason, it is a good idea to read all of the books on competition driving and study the driving instruction tapes, and also to hook up with an established driver who can be your mentor and help you learn to drive.

It is also advisable to rent some track time prior to your school, and get some on-track experience so that your mentor has an opportunity to watch you drive and help you correct any mistakes you are making. Going to the trouble of doing this will make you look much better at your school and will provide you with a better chance of having the second school waived. Even if you have previous experience in another type of racing, this may be a good idea to get to know the type of car you will be driving.

The school will begin on Friday evening, with a classroom session usually held in a meeting room of the race headquarters hotel. In the classroom, you will meet the individual instructors and the Chief Instructor. They will explain a little about how to pick lines and will answer the students' questions on anything from shifting technique to adjusting tire pressure. A large portion of the evening, however, will be spent discussing the function of the Stewards of the Meet, what to do when an emergency vehicle is on the track, and the meaning of the flags. This is all essential knowledge, but can be learned from reading the GCR. You should own a copy of the current GCR and the appropriate specification book for your category of car before running your first school. You should also be fluent in the rules presented in them.

On Saturday, you will meet with your particular instructor at the track, and he will first drive you and your fellow students around the track in a street car while explaining the lines he uses in his race car. After a few laps with him, you will strap into your own race car and go out on the track behind a pace car. No passing is allowed during these exploratory laps. After this session, your instructor will talk to you and answer any questions that have come up. A second session will then be run with full-course green flags, so that you can run at racing speeds when passing is allowed. After this session, the instructor will once again answer questions and this time will critique your driving. This sequence of driving and critiquing will continue all day. The sessions are usually twenty minutes long, with not more than thirty minutes in between, so it is imperative that your car be reliable and that you have plenty of good mechanical help to repair anything that goes wrong with it between sessions. You will not have time to do anything on it yourself.

At some point(s) during the day, the flagman will show you various other flags besides the green and checkered, to make sure that you know their meanings. You may be shown a black flag directed only at you or a red flag for the entire field. Besides learning the meanings of the flags, another lesson will become apparent: expect the unexpected. Readiness to react to an unexpected problem or situation will be required in almost every race you ever run.

In the final session, a couple of trial starts will be conducted, which will be races to the first corner. At this point, learning to anticipate the flag and "outdrag" your competitors is important, but not essential. This session will be concluded by a short race of about three laps as sort of a summary of what you have learned during the day. No one expects you to win the school race and no one will care if you do. Just make sure that you look like you know what you are doing.

At the conclusion of the day, your instructor will again critique your driving, and will score you on a number of characteristics including: attitude, consistency, courtesy, judgement, technique, reactions, flag recognition, and knowledge of the GCR. Strive for the best score possible in each of these. Again, good scores will help you to get your second school waived, so that you can run the Regional race the following day.

Registration

Once you are an SCCA member and have a Competition License or a Novice Permit, you will receive in the mail an entry form for each race in your Division. If you plan to go to the race, fill it out and mail it in early, with the entry fee. Mailing your entry in on Tuesday before the race and finding out on Saturday morning at the track that the Registrar has not yet received it just leads to frustration for both of you.

Once at the track on Saturday morning, the Registrar should have your entry on file. To complete the registration process, you will need to show your credentials (your membership card and license or permit) and sign a release of liability. Each of your crew members must do the same. If you need to add crew members for the weekend, this is the time to add them to the list. Guests and sponsors who will be joining you during the weekend should be put on the list as over crew. You will receive a paddock pass, which should be displayed on the windshield of your tow vehicle for the weekend. Display your license on your person during the weekend, and have each of your crew members do the same. Licenses will be required for entrance to the paddock and the hot pit area. *DON'T LOSE THEM!* In reality, officials do not usually bother to check for credentials on those wearing driver's suits in the paddock.

Schedule

The schedule for the weekend's events is divided into race groups, which are categories of similar type and speed cars run together on the track. Announcements calling your group to the grid will be made by calling the race group number, so you must know which one includes your class. The schedule will not give specific times for each group to go onto the track. This is done because it is very difficult for those running the event to adhere to a set schedule by time. Instead, it is up to you to know how long each group is on the track and how many groups run before you. By keeping track of which group is running, you will know approximately how much time is available to make the repairs to the car that inevitably arise, before you must leave for the grid.

Listen carefully to announcements made over the PA system. It is much too easy to fall into a trap of not paying attention to them because many of them are unimportant to you. Occasionally, however, an announcement will be made that is important, such as a shortening of each race group's on-track time for the remainder of the day. Missing an announcement like this can cause you to miss an on track session, something that should be avoided like the plague. It is beneficial to be the first one in line on the grid for a session. Not only are you assured of a track clear of oil or debris for one lap, but it also makes your team look good and is a morale-builder for the crew.

In the Paddock

Paddock etiquette is mostly a matter of common sense. SCCA's GCR says that a driver is responsible for the action of his crew in the paddock, since he is the one who has the most to lose as a result of the improper behavior of one in his charge. The Club can pull the driver's license. In practice, in almost 25 years of competition with SCCA, I have never seen anything happen other than a verbal reprimand. Offenses may include drunken or disorderly conduct and fighting, but most often are arguments with officials. This is not to say that no one ever cuts up in the paddock, throws a frisbee, or has a beer. When something like this causes a problem for another competitor, however, expect a visit from the Chief Steward.

It should be pointed out that alcoholic beverages are strictly prohibited in the paddock and pit area until the last car has come in from the track for the day. Many competitors and their crew will have a beer after their last session of the day, though, even though other sessions are still on the track. If you partake or allow your crew to do so, make sure the evidence is disguised in cups or in soft drink cans. Failure to do so may get you thrown out. I allow the crew to have a couple of beers at the end of the day, but the driver is under strict orders not to consume anything alcoholic until after his race at the conclusion of the weekend. This policy has caused me to be unpopular with some of my drivers on occasion; however, if they are serious, they know that this is the price of success.

Other forms of paddock etiquette are common sense, also. A racer must basically work and live with his competitors for a weekend. This requires thinking of your neighbor's needs and desires. Although forcing him to listen to your generator may be unavoidable, being a burden to him is not. Racers help each other a lot. Borrowing tools and supplies goes on all the time, but if you are ill prepared, try to spread out the burden. Don't depend on any one team to solve all of your problems. When you do use parts, supplies, or materials provided by someone else, always pay for what you used. Even if it is just a quart of oil or a brazing rod, give that person a couple of dollars. Think of how you would feel in his situation. If you borrow a tool, return it as soon as you are finished with it. Don't wait until it is convenient. If you loan someone a hacksaw, and find two hours later that you need it yourself, but don't have it, how would you feel?

One successful racer has a very well-equipped truck that includes a TIG welder, although he is not primarily a fabricator. Since it is the only one at many of the races I attend, I seem to be borrowing it frequently to help one of my customers or another racer. In exchange, I have helped the owner of the equipment learn how to use it better. My client or the racer, who is the primary beneficiary of the welding, usually gives him a few bucks to pay for gasoline, argon, and welding rod. What goes around, comes around.

While you are completing the requirements of a Novice Permit, don't forget that the Chief Steward must sign it after your race is complete. Failure to have him do this will result in loss of credit for that race for license requirements. If your car is involved in a serious accident, you can expect a visit from the Chief Steward and/or the Tech Inspector. The incident will be written up in the Vehicle Log Book as a running history of what has happened to the car.

It is also a good idea to keep both the qualifying and race results sheets from Timing and Scoring. This will give you documentation of your fastest laps, laps in traffic, finishing position, etc. You never know when these might be necessary. There have been instances where a driver could not renew his National License because Denver claimed he had not completed the two-race minimum for a given year. Having the official race results sheets refuted this claim. When it comes time to sell your car or to try to attract a sponsor, your official results may be beneficial, too.

Since most everyone who club races must be at work at 8 AM on Monday, there is always a mad rush for the gate as soon as each race is over. Loading is usually hurried, and the truck or trailer is not always packed as carefully for the trip home as it was in preparation for the race. At some tracks where the paddock is in the infield, racers find a hurried loading job to be followed by up to 30 minutes of waiting for a race to end before they can cross the track to leave. Thus, planning and foresight are important here.

The time after a race is a time for celebration by a few and reflection by all. Immediately after a race is the best time for the entire crew to go over what has happened during the weekend. The event and all of its parts should be critiqued to discover how the team could do better next time. Even if you won, it is likely that some sort of a minor catastrophe occurred which caused concern or additional work. Sometimes it is after a race that it becomes evident that a better strategy would have worked better or that a crew member made a mistake. But the purpose of this debriefing session is not to reprimand anyone. If a mistake was made, the entire crew already knows it and placing blame will only make the perpetrator feel worse. The goal is to learn from the mistake, so that no one in the team will make the same mistake again.

This is also a good time for the driver and engineer to confer on the events of the weekend as they relate to chassis setup. During the weekend, there is usually only time for discussing setup for the next session. After the race, though, the entire weekend can be thought through in hindsight, and the possibility is great that better ways can be conceived to solve the problems that came up. These post-race debriefings will also generate new ideas for setups to try in testing later. These discussions sometimes go on for seven or eight hours before the participants tire of them or arrive home.

In addition to all of the other benefits of an after-race bull session, the driver can finally have a beer—hardly ample reward for all he has been through during the weekend—but even then, only if some one else is driving the tow vehicle.

Protests

After you have your National License, your contact with SCCA officials will be more limited; however, when contact with officials is required at this point in your career, it may be of a more serious nature. No one enjoys being involved in protests, but in any competitive activity, disputes are inevitable. As in the federal government, there are two types of legal actions. If another competitor protests you for technical illegality or for an on-track driving rules infraction, it works like a civil case. If the Club proceeds against you, however, it works more like a criminal case.

A protest may be filed by any club member who is directly connected with a specific race. This includes drivers, entrants, and crew entered in a race, as well as SCCA officials. Someone at a race meeting who is not involved with the race in question cannot file a protest. A GT driver cannot protest an F/F competitor, for instance. An official can protest a competitor, and a competitor can protest an official. To file a protest, the protestor must pay a $50 fee and explain, in writing, who committed the alleged offense and exactly what part of the GCR or Specification book was compromised. If a competitor is being protested for a technical violation, a teardown bond is also required. If the protestee is found to be legal, he is awarded the bond for reassembly.

The Stewards of the Meet (SOM) will hear evidence, in person, by phone, or in documentation before they reach a decision. After their decision has been made, any party connected with the event has the right to appeal the decision. The Appeals Board then repeats the procedure and comes to its own decision as to guilt or innocence, legality, and/or penalty. Their decision is completely separate from that of the SOM and is binding as the final authority.

If you intend to file a protest, you should have a valid reason backed up by the GCR or the appropriate Specification book. You should also back the protest up with any documentation or witness testimony that you can find. To file a protest, see a Steward of the Meet, fill out the form he gives you, and pay him the appropriate fee.

If you have been protested, first ask yourself if the action is justified. If a competitor smelled nitromethane in your exhaust while following you, and you were using an illegal fuel, shame on you. Take your lumps, and don't do it again. If, on the other hand, you know you did nothing illegal or if you found a loophole in the rules, gather all the evidence you can to support your claim and go in with the knowledge that you are right. If the protest is filed by a competitor, it is up to him to prove that you were in the wrong. If indeed you are wrong, this may be relatively easy. If SCCA brings an action against you, though, whether a disqualification, an on-track rules infringement, or a technical infraction, it is up to you to prove them wrong. This is usually an uphill battle.

Protests often involve a driving violation of the "rules of the road." For example, against a driver who passes another in a corner where a yellow flag is being shown. In this example, the overtaken driver files a protest on the passing driver, and the corner workers are interviewed to find out what they saw. The Stewards of the Meet rule on the protest and find the passing driver either guilty or not guilty. Depending on the severity of the infraction and the personality and mood of the Stewards, the penalty can be anything from a time penalty to a loss of the driver's license.

Protests between competitors can also involve technical matters. Back in the 1970s, one Formula Ford driver knew that his engine builder was using an illegal head on his own car. The driver purchased an identical head from the engine builder for his car, but did not use it. When the engine builder finished second at the Runoffs™ that year, the driver filed a protest on the engine builder for using an illegal head and offered his own as evidence! The engine builder was penalized with a six-month license suspension. This whole incident was so underhanded that you really have to wonder what had happened between the driver and the engine builder before.

Protests can be filed by a competitor on the actions of the Club, too, as was the case at the Runoffs™ in 1991. Dave Salls won the F Production race in an MG Midget. In the post-race teardown, the Tech Inspector found Salls illegal for having one intake valve .0065 of an inch too large and another .0070 too large. The Production Car Specifications at that time stated that the valve size was to be 1.31 inches. I was Dave's engineer that year, and we thought he was completely legal. The initial penalty was assessed at $2500. Within 15 minutes it was lowered to $1,000. (I still do not understand the justification for monetary penalties in amateur racing.) Thinking that, if indeed the valves were too big and the micrometer doesn't lie, we made a mistake by filing a protest against SCCA for the severity of the penalty rather than the illegality ruling. When the case came before the Stewards of the Meet, people came out of the woodwork to testify for Salls. The late Dave Taber (a fine individual if there ever was one) was the engine builder for a number of other drivers in the race, but did not build Salls's engine. He and several competitors who finished behind Salls testified on his behalf. Most of the people involved in SCCA racing are great! Everyone, including the Stewards, agreed that even if the valves were too large, seven thousands of an inch is not a performance advantage. After much discussion, the SOMs decided in Salls's favor and reinstated the victory. They accepted Taber's argument that a specification spelled out to two decimal places couldn't be measured to four. In light of the fact that the penalty had been protested and not the illegality, this was a major relief.

The Supplementary Regulations for the Runoffs™ stated that an appeal of a protest decision by the SOMs must be filed within one hour of the posting of the decision. We sat on beds of nails for that hour, but when the time was up, we thought we were home free. A short time later, however, Dave overheard a conversation between another F/P competitor and a high ranking official. The official asked, "How did you finish?" When the competitor told him, the official replied, "Well just wait. We aren't done yet." Sometimes it seems as if there is less politics in Washington than in racing. Almost three hours after the finding of the SOMs in Salls's favor, an appeal was filed by the Chief Steward. That signified that the end was near. When Dave went before the Appeals Board, they told him they had found against him. When he asked if he could present his case, they said, "Of course. By all means." He presented his side of the story and they said, "That's all very good, but we still find against you." The next year the rule that disqualified Salls was changed.

When you are racing competitively, protests are a fact of life. Experience in protest situations helps, as does experience in dealing with different personality types and the politics involved when officials are engaged in power struggles. We learned a great deal from this episode, but at the price of a National Championship. It is doubtful that anything could have been done to alter the outcome; it had much less to do with what happened at Atlanta than it did with what was going on in Denver. Some days it seems you just can't win. Dave quit racing as a result of this incident, and that is to SCCA's and the sport's detriment. To be successful, though, you cannot let these things get you down for long. In Dave's case, since he quit, he let SCCA win.

Dave Salls took the checkered flag first, but the race with officials in the tech barn proved to be a bit tougher. That's the author in the middle of the top row; Dave, in the driver's seat is almost obscured.

Chapter 5

Dealing with Risk

I get scared almost every time I get in a race car. And, if I'm not scaring myself, someone else is doing it for me. I'm not like most of the other guys who say they've never been scared in a race car. Well, if they haven't been, they're just not driving as hard as they should be. Because if you drive a race car as hard as it will go, you're bound to scare yourself.—A.J. Foyt

The two groups of drivers, those who do it as a hobby and those who make a career of it, have very different outlooks on the dangers of racing. A driver who races for fun probably has a family, a job, and all of the attendant responsibilities. Thus, even if he is very serious about his hobby, he must still make sure that he can provide for his family and take care of his responsibilities. As a result, such a driver tends to take a somewhat more conservative approach than a driver who races for a living. Either way, nobody wants to get hurt. It is important to realize, though, that pushing the envelope is risky business.

Most amateurs race for their entire careers without being injured. Ask someone who has been running SCCA club races for two decades and he will probably tell you about dozens of incidents he has been involved in, but he will also probably say that he has never been hurt. In part, this is due to the limited track time that club racing provides. Pros, however, run more than the two hours you see on TV every Sunday afternoon—*a lot more*. Professional drivers may (or may not) have more innate talent than club racers, but they do have one big advantage—track time. As with any another learned skill, practice makes perfect. The more you can practice, the closer to perfect you can become, but with this advantage comes a real liability. Track time provides more opportunity for off-course excursions.

Another liability for pros is that they are under a great deal of pressure. They have the same drive to win as the club racers, but they are being paid to go fast, so there is more pressure to perform. Sponsors like to see their cars come home first. This means that they will charge just a little harder. This is especially true during qualifying when some laps at ten tenths must be put in. Driving closer to the limit increases risk. If you intend to race for a living, you had better accept the fact now that, sooner or later, you will get hurt. Although it is something that none of us enjoys, hospital time is a fact of life for a professional driver.

Some years ago, Dr. Michael Henderson, a medical doctor and racing enthusiast, analyzed 221 racing accidents that occurred in England. His findings are particularly interesting. One of his criteria for calling an incident an "accident" was that the car was not able to be repaired to racing condition within 24 hours. About the cars involved in this study, Dr. Henderson explains, "In this survey, the incidence of crashes was roughly proportional to the relative popularity of the classes, but small saloon cars had a very high number of minor incidents (outside the terms of reference of the investigation). Single-seater racing cars were involved in 15% of the accidents investigated, open two-seater sports cars in 17%, closed two-seater sports cars in 25%, and saloon cars (of a nominal four seats) in 40% of the total number." In this country, we do not really have a classification for closed two-seater sports cars, but the GT category includes some of these cars, as well as many of the "saloon cars." Of the 221 accidents, Dr. Henderson determined that 77% were a result of driver error, either the error of the driver of the car involved or that of another driver, 14% resulted from mechanical failure, and 9% were attributed to "other causes."

Dr. Henderson also showed that closed cars are more likely to overturn in an accident, with 35% of the accidents which involved closed cars ending on the roof. Formula and open sports cars were almost equally split at 25% and 24% respectively.

Of the 221 accidents, single-seater drivers were injured in 38% of the crashes in which they were involved, open sports car drivers in 37%, closed sports car drivers in 17%, and saloon car drivers in 19%. If we combine the statistics for closed sports and saloon cars, the likelihood of injury in closed cars appears to be 36%, showing that no one of these categories is inherently more dangerous than the others. It is regrettable that Dr. Henderson's statistics on severity of injury are a bit vague and difficult to relate to present-day racing. It would be interesting to separate sprained thumbs from more serious injuries. To the best of my knowledge, this is the only study of this type ever conducted.

Safety equipment has come a long way since Henderson's survey, and it is safe to assume that the incidence of injury has decreased in recent years. On the other hand, with tires being substantially stickier than they used to be, impacts are now much greater when two cars collide because the tires do not slide as easily.

Be advised, whether you are running an HP Bugeye or an Indy car, that the potential for injury is real. Foyt says, "If you want to know, every car is involved in four or five near misses in every race. It doesn't matter if it's Indy or some little old dirt track." In any risky activity, precautions must be taken to minimize the danger. Pilots have parachutes. We have roll cages, helmets, fire resistant suits, safety seats and belts, fire extinguishers, and all the rest. They do a very good job, and, happily, accidents resulting in injury are not nearly as common as they were in Dr. Henderson's time.

Safety Equipment

It can happen to you—and then it can happen to you again.—Steve McQueen in *Le Mans*

The probability of accidents should not be cause for alarm. Equipment is available to prevent or reduce the possibility of injuries, and you should take full advantage of it. In racing, there are generally two types of injuries: those caused by impacts and burns. Most injuries are caused by impacts.

Head and Neck Protection

The main function of the helmet is to protect the driver's head from impacting foreign objects in a crash. To do this, the outer shell must have great structural integrity. For many years, the shells of good helmets were made from fiberglass, and many are still made from it today. With advances in technology, though, today's best are made from a combination of Kevlar™ and carbon fiber. Kevlar™ is an aramid fiber that is extremely strong and tough. The bulletproof vests in use by the police and military are made from it, so it should be apparent that it is very resistant to puncture and tearing. When Kevlar™ is laminated with carbon fiber, which is very stiff, it makes for a very hard, tough shell. And yes, it is also very light. Many SCCA club racing cars are capable of generating cornering forces of up to two and a half Gs. A helmet that weighs a mere three pounds more than one made from Kevlar™/carbon will feel like it weighs seven and a half pounds more when cornering at this level.

Inside the shell is a compressible liner that is designed to crush and thus absorb energy when the shell forcibly contacts an object. This allows the head to slow gradually during an impact and will lessen the possibility of concussion or other severe injury. All else being equal, the thicker the liner, the more energy will be dissipated, and the more gradually the head will come to a stop.

Comfort padding in the helmet is covered with either a nylon fabric, in the cheaper models, or Nomex™, in the more expensive ones. Most helmets made for racing use padding placed firmly against the cheekbones, just below the eyes to hold the helmet in place on your face. The last thing you want is a helmet that slops around on your head. It should be a snug fit without being tight. For this reason, you should buy your helmet in person, so you can try it on. Do not buy by mail. The chinstrap should be long enough that you do not have trouble either fastening or loosening it. Remember, when you do this, you will sometimes be belted into the car in cramped quarters. Fastening or loosening the chinstrap can be much more difficult than you might expect, so make sure it is easy to do in the store. I once saw a driver use a quick-release mechanism on the chinstrap of his helmet. A few weeks later, he had an accident and the latch came undone and the helmet came off of his head. Fortunately, he has a hard head and was not injured.

The last piece of mandatory equipment on the helmet is the face shield. Not too many years ago all shields were made from a polycarbonate material, which was quite brittle. Some drivers lost eyes when hit in the visor with rocks in open cars. No driver in his right mind would now use anything other than a Lexan shield, preferably at least 1/8 inch thick. Incidentally, it is a good idea to use both a full coverage helmet and a face shield in a closed car, especially if you are running with an organization that still requires a glass windshield. If the windshield is broken, either from rocks thrown from other cars or in a crash, the glass will splinter and go everywhere, including in your eyes if they are not protected.

Many helmets are sold now with bells and whistles that do little to enhance the basic safety functions, but can still be beneficial. The best of these are the ventilation systems which help prevent fogging of the visor during rain or humid conditions. Some of these work better than others, so talk to users of the helmets you are considering before you buy. One helmet has "vortex generators" on its front and side surfaces, supposedly to reduce drag and buffeting at speed in open cars. Unless you are running at Indy, this is probably only flash. Other helmets have larger or smaller than normal eye openings. The smaller the opening, the more area is protected, but choose the one that feels right. It is better to be comfortable than to have an extra square inch of shell. A very beneficial option on many high-end helmets is a fitting for introducing breathing air into it from a pressurized bottle. In a fire, the oxygen in and around the car will be consumed, leaving little for the driver until he is removed from the car. A supply of air piped into the helmet will give the driver a minute or more of air during this time.

The Snell Foundation grades helmets on safety and groups them by suitability. The helmets that they approve for use receive a "Snell sticker," which is required by SCCA and most other sanctioning bodies. Competition helmets carry an SA designation such as SA95, indicating that it was tested in 1995. A similar motorcycle helmet would carry an M95 designation, and is not approved for automotive competition. The Snell Foundation tests helmets every five years and the sanctioning bodies only allow helmets to be used which bear the latest Snell sticker. This means your helmet must be replaced every five years.

Regardless of which helmet you choose, what it is made from, and what it's features are, it will wear out. Two factors are important here: the sweat and oils absorbed by the liner and the shock that the shell receives. Sweat and oils absorbed by the liner will break down the composition of the material and reduce its energy absorption value. Shocks to the shell occur when the helmet is bumped and banged into things or is dropped. (The driver who throws his helmet after having the door shut by a competitor or losing an important race has no place in racing. He is too emotional, and is probably using a damaged helmet!) If you are racing only seven or eight times a year, your helmet should be replaced every two years. If you get to run more frequently, annual replacement is best. An alternative is to send your helmet back to the manufacturer to have the liner replaced. Any time the helmet experiences a severe blow it should be inspected by the manufacturer as well. They usually only charge a modest fee for this service.

Attention in recent years has focused on preventing the neck injuries that occur as a result of a serious accident because of the forces acting on the driver. Although the belts restrain a driver's torso and the helmet protects his head, these are not adequate means of protecting the driver's neck when high G forces work on the weight of his head and helmet to put undue strain on his

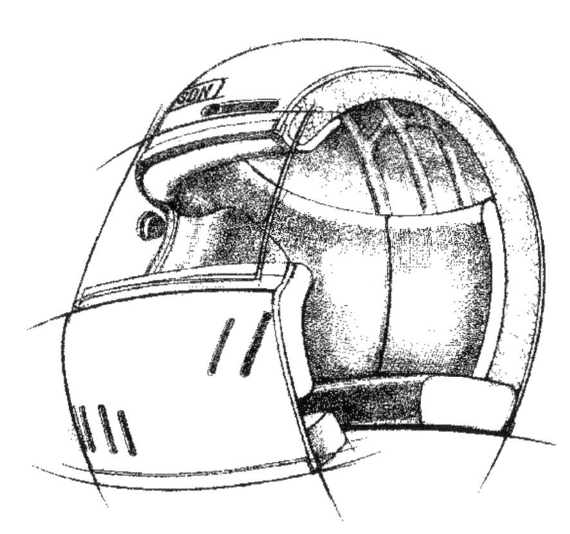

Fig. 5-1 This helmet cutaway view shows an outer layer of fiberglass or carbon/Kevlar™, a thick middle layer of a crushable, energy-absorbant foam, and an inner layer of comfort padding. Illustration courtesy of Simpson Products.

spine. A device has been produced by Hubbard-Downing racing called the Helmet and Neck Support or HANS system. It uses a composite yoke, which is retained by the shoulder straps, which provides a sort of built-in head rest to restrain the helmet laterally and to the rear. Small straps attached to the helmet keep the head from moving violently to the front. The system seems quite effective, and some of its users credit it with their ability to return to race again after serious crashes. The only drawback of the HANS system seems to be its restrictive nature in confined cockpits.

Another method that is being tested in Europe is an airbag system for helmets. With this system, a small donut-shaped device attaches to the bottom of the helmet and is deployed downward when a sensor mounted in the car detects an abnormally high G load. Like airbags in street automobiles, this happens very quickly, and the airbag supports the head to prevent movement in any direction. Although, this system is not yet commercially available, its disadvantage is sure to be cost when it is released for use.

After the devastating accidents in Formula One in 1994, FOCA adopted a mandatory headrest for its cars. This is made from a crushable foam three inches thick behind the helmet, and wraps around to protect the head laterally as well. This is certainly a step in the right direction. Such a headrest, combined with a good helmet and a helmet airbag or the HANS system, might go a long way toward preventing neck and head injuries.

The Seat/Belt System

Although it may not be readily apparent, the seat is a very important piece of safety equipment. Along with the belts, the seat is responsible for holding the driver in place, both in normal operation and in the event of a shunt. The seat and the belts should be thought of as the seat/belt system. When the belts are tightened, they pull the driver down against and into the seat. It should be obvious that, if the seat or its mounts bend or break during an accident, the belts will do little good. The seat, then, is the foundation of the system.

In a production-based car, the seat is generally produced by a commercial manufacturer, and should be one intended for racing. As with helmets, cheaper seats are made from fiberglass, and more expensive ones from carbon and/or Kevlar™. Fiberglass or plastic seats built for dune buggies are dangerous when used in racing. Many oval track-sanctioning bodies will not allow fiberglass seats at all. Since anyone with a mold can lay up a glass part, some inferior ones turned up in oval track cars and hurt their drivers. They now require aluminum ones, which usually have a "wrap around" section under the arm on the right side. Some seats have been produced with a "double wrap," and have been used in road racing cars. Doug Wolfgang, a top World of Outlaws driver, crashed in 1993 and was trapped in a burning car. Doug was unconscious and rescue workers had trouble removing him from the car, in part, because his torso had to be turned sideways to get him out of the double wrap seat. This could be especially dangerous if a spinal injury is suspected.

Select a seat with high sides at the hips and good lateral support at the shoulders. It should be stiff and have mounts that are molded in, not just bolted on with hardware store fasteners. The padding should be firm, not too cushy, and the seat should fit your body. The padding should be covered with cloth, not vinyl, to allow it to breathe. Vinyl gets hot in the sun and wet when a sweaty driver's suit is seated against it. A dead giveaway, that a seat was not intended for racing or is otherwise less desirable, is if it does not have slots for the belts to pass through, or if the slots are not properly finished out with edge protectors.

Many otherwise respectable Formula and Sports Racing car constructors provide the cheapest seats known in the free world. One very fine driver with whom I worked had to move the glass seat that came in his Swift forward and up slightly to fit the car to him. He did this by using a light aluminum bracket, which he attached to the top of the seat with AN hardware. Sometime later, he backed the car into a wall, and the aluminum bracket bent and the fiberglass broke where the bracket was attached. Although damage to the car was not extensive, he suffered a spinal injury which kept him from racing for two years. A proper seat, properly mounted, could have prevented this.

In cars of this type, where sheet metal panels form a "bucket" where the driver sits, a popular and very effective seat is made with what is known as "two part foam." To build a seat, a double layer of trash bags is laid into the floor of the car, and the driver sits on top of them. Then, foam is poured into the trash bags. In its original state, this foam is two different liquids. When they are poured together and mixed, they begin to turn into a liquid foam, which is poured into the trash bags. After about 15 minutes, the liquid will solidify and make a very firm, dense, yet light, seat that fits the driver exactly. The foam seat should be taped with racer's tape, so that small bits will not break off. It may then be covered with any appropriate fabric. The major drawback of a foam seat is that the solid foam is actually a polyurethane. As safety-conscious air travelers know, this plastic gives off toxic gases when it burns. Some consolation may be derived from the fact that, if the seat burns, toxic gas is going to be pretty low on your priority list. A relatively new seat material is the "Indi Seat." This is made in a similar manner to the two part foam seat, but uses plastic beads poured into a plastic bag from which the air is then removed with a vacuum pump. Studies suggest that this seat is better at absorbing energy than the two-part seat and may be safer.

Belts should be made for competition and have at least five attachment points, although six is better. The "lever latch" belts that were popular for a long time are, fortunately, on their way out. Because they require so many parts of the belt assembly to be held together at the same time while the lever is latched, they are extremely cumbersome and definitely not suited for tight cockpits. Cam lock belts are becoming widespread. Some of the camlock latches are better than others, however. Some have handles so small that they are difficult to grasp with gloved hands, and others tend to open when arm restraint straps hang on them. It is quite disconcerting to a driver to have his belts suddenly unlatched halfway through a corner. Besides, that means a stop at his pit to have them relatched. Good belts are made by Simpson, TRW Sabelt, Willans, and Luke.

All of these manufacturers make belts with three different attachment systems: bolt in, tube mount, and clip on. The clip ons have the advantage of being able to be removed when washing the car. Their disadvantage is the unlikely possibility that they may become unclipped during use. Tube mounted belts need something on each side of each strap at the mounting point to keep them from sliding. Bolted in belts are definitely the most secure, but the lap belt mountings are usually hidden down under the seat where they get little attention. The type of mounting system required depends on your car and the type of maintenance it receives.

Regardless of the mounting system used, common engineering sense is required. Once, when helping to take apart a Sprint Car which had been crashed heavily, I discovered a bolt in lap belt mounting tab, which had originally not lined up properly, and had been heated and bent about 45 degrees. The mounting tabs on belts are made from special alloys that have been carefully heat-treated. When heated and/or bent, the metallurgy is affected. This particular tab had cracked about two thirds across as a result of the forces on the belts during the crash. The driver almost came out of the car. Make sure the mounts on the chassis correctly mate with the tabs on the belts, if they are of the bolt in style. Do not modify the tabs that come on the belts. Like helmets, belts should also be replaced every two years.

Besides using quality belts and making sure they are in good condition and properly connected to the chassis, it is also important to remember to tighten the belts properly. Forces on the driver in a crash can exceed ten Gs momentarily, and a 180-pound driver at ten Gs equals a force of 1800 pounds distributed among the belts. These forces almost always stretch the belts a great deal. The stretch of the belts helps to keep the instantaneous forces on the driver down, but it can also allow the driver enough movement to contact objects inside or even outside the car. *Open Wheel* magazine regularly publishes photos of Sprint Car drivers upside-down in midair with arms and head outside of the roll cage. Belts stretch—a lot! Keep them tight and you will have a better chance of staying inside the roll cage in a similar situation.

If the helmet protects the head and the seat/ belt system keeps the body inside the roll cage in the event of an accident, consideration must still be given to preventing impact injuries that occur when the driver's limbs contact parts of the driver's compartment. The most common means of accomplishing this is to pad anything that can be hit. When I was racing, I rarely finished a race without a bruise somewhere, and I tried to pad *everything*. It is amazing what your knees, elbows, shoulders, and ankles can find in a cockpit to hit. Even more amazing is what will hit the body in the event of a crash. Parts of the car that are normally impossible to contact impail themselves into your body in a violent wreck. Be safety conscious. Try to eliminate *every* possibility.

Arm restraints are required in all open cars to prevent the driver from reaching his arms or hands outside of the cage. In order to be effective, arm restraints should be attached to the forearms. Fastening them at the elbows or higher allows considerably more movement and negates the proper function of the restraints. To keep them from slipping, some drivers sew the armbands to which the straps attach to their suit.

The Roll Cage

The best seat, belts, arm restraints, and helmet are of no use whatsoever if the roll cage is not up to the task. SCCA, as well as most other sanctioning bodies, have very good roll cage specifications for all cars—except Showroom Stock and A Sedan. Denver requires cages for these cars to be bolted in to the chassis and bolted together so that the cage will not impart any additional stiffness to the car. This is a stupid rule. Who cares if the chassis is a little stiffer, as long as the driver is properly protected? I have seen two of these cages fold up in Showroom Stockers, one in a Camaro, and the other in a Mazda. The Camaro driver was lucky; the Mazda driver, Jay Wright, was not. This is the primary reason I don't recommend Showroom Stock racing.

In a letter to *Sports Car* Magazine in October of 1994 Jay said, "A friend of mine worked with the head of NHRA safety and they had my cage evaluated by an aircraft designer. The results from this led to NHRA requiring the addition of three extra bars in their Street Stock category. One of which—a straight bar running across the main hoop and through the diagonal—can be shown that it would have prevented my neck being broken. The fact that modification of the cage could have prevented any part of my injuries justifies, to me, allowing better cages regardless of the so called advantages they may produce." As of the date of Jay's letter, SCCA had still not allowed proper roll cages in Showroom Stockers.

As late as midyear 1995, a letter from another member was printed in the Club's publication, which asked the Comp Board to increase the safety of cages and seats in A Sedan. In response, the Comp Board Chairman tried to justify their tardiness by saying, "The Comp Board is currently proposing changes to the AS roll cage rules. While not totally devoid of risk, Club Racing is not a contact sport." NHRA obviously has less inertia when it comes to improving things.

In 1983 Rolf Stommelen drove a Porsche 935 in the LA Times six-hour endurance race at Riverside. Porsche builds strong, advanced engines and great gearboxes, but they have traditionally been a little behind the times with regard to chassis technology. Stommelen's car had been built with an *aluminum* roll cage! This is something they had been doing for years. I remarked at the time, without even knowing who was running the car, that I would not want to drive it. Toward the end of the race, Stommelen hit the outside wall in Turn 9 and cartwheeled almost up to the pit entrance. When the car stopped, it was not even recognizable as a race car. The cage came apart and Stommelen was killed. Roll cages are important. Don't be tempted to cut corners for weight reduction, and make sure the welding is done by someone who *really* knows how to do it. IMSA outlawed aluminum cages the next year.

Fire Protection

Fire is a rather rare occurrence in racing, in part due to the high-quality fuel cells in use today. However, because fire is so devastating when it does occur, a great deal of attention is afforded to fire prevention. All sanctioning bodies have minimum requirements for fuel cells, and all are very good. Formula Cars are designed to lose the engine/gearbox/rear suspension in the event of a major crash. Since the fuel cell(s) usually stays with the chassis, CART and IRL now require an automatically sealing valve on the hose from the fuel cell to prevent spillage if the car is broken in half. This valve is a worthwhile addition to any Formula or Sports Racing Car. Front-engined cars usually have the fuel cell located in the trunk area. The fuel cell should be located as far forward as possible, mounted on good support structure (not just the trunk floor sheet metal), and be protected with crushable structure behind it. The job of the fuel cell is to contain the fuel in the event of a crash. Make it easy for it to do its job by removing the objects that could puncture it.

The next line of defense against fire is an effective extinguisher system. In SCCA, all cars are required to be equipped with some type of extinguisher. Regardless of the rules, a multi-point halon or equivalent system should be employed. As this is written, the EPA has cracked down on all chlorofluorocarbon compounds from those in spray paint to halon. New systems are now being tested as halon replacements, but as yet, none has emerged as a clear leader.

Just as every accident is different, so is every fire. Some are fueled by gasoline, some by alcohol, some by fiberglass and rubber, and some by magnesium. Every firefighter knows that each fire requires specific extinguishing methods, so one extinguisher medium will not be optimum for all conditions. Generally, however, fuel will ignite first, then normal combustibles such as oil, fiberglass, paint, plastic, rubber, etc., and then magnesium and titanium. An extinguisher system, since it is only functional for a minute or less, should be directed toward the first compound to ignite: fuel. By the time the extinguisher is exhausted, if the fire has not been put out, the combustibles will begin to ignite. It is logical, therefore, to use the largest extinguisher system available that will fit the chassis. This will give the driver the best chance of escaping without burn injuries.

If the first line of defense in fire protection is to prevent fuel spillage, and the second is to extinguish the fire, then the third is burn prevention. This is the job of the driver's suit. Most suits are made from a Du Pont fabric called Nomex™, although there are many others including Proban™, FPT™, and Kevlar™. To protect the driver's skin from burns, the suit must perform two

The possibility of injury in racing is real, and the risks must not be taken lightly. Safety items such as a good-quality driver's suit, fire-resistant balaclava (simply called a "hood" by most people), a good helmet with medical information on the back, and a well-designed and -constructed roll cage may keep you from meeting the doctors and nurses.

functions. The outer layer of the suit must remain intact and flexible, even when exposed to flame for up to a minute. The inner layers must insulate the skin from the high temperatures of the outer layer during this time. The more insulating layers included in the suit, the more time the driver has before the temperature inside the suit rises to the point of burning the skin. Quilted suits have the best insulation properties. SCCA requires the use of fire-resistant long underwear with any suit of less than three layers. This is a good idea even with a quilted suit.

Gloves and shoes must meet the same outer layer integrity and insulation requirements as the suit, but must also allow a firm grip on the wheel and pedals without slipping. Suede leather has become popular for shoe construction, and most gloves use leather in the palms. Although leather has some very good properties of durability, good looks, and grip, it shrinks when heated or burned. In a fire, leather shoes may shrink around the driver's feet, increasing heat transfer to the skin and making removal of the shoes difficult. Better shoes would be those constructed from a fire-resistant material such as Nomex™. Shoes should be high-topped, and gloves should be of a gauntlet style that will overlap substantially with sleeves.

Deceleration Forces

According to paddock wisdom, the only danger in a crash is the sudden stop. This truism is so obvious as to be absurd, but it is really very accurate in describing what actually happens in a crash in relation to the limits of the human body. Many wrecks on road courses involve cars leaving the track and traveling some distance, usually while spinning, before striking some immovable object at some angle other than head-on. As we will explore in a moment, the dissipation of a large amount of kinetic energy within a short span of time does the most damage to the body. During one of these road course crashes, a great deal of energy is dissipated when the tires are scrubbing through dirt and turf. If enough runoff area is available to sufficiently reduce the energy level before the impact, little damage will be done to the car and driver. Gravel pits have proven very successful in this respect, but make removal of the car quite difficult.

The human body is a rather delicate and fragile biological assembly. It has adapted during millions of years to be suitable for walking upright, handling small objects, and moving at no more than running speeds. It was never intended to handle G forces produced from cornering or braking in a modern race car, and is certainly not suited to endure the G forces produced in an accident. The basic problem seems to be the inability of the brain, heart, and other organs to function or even remain in their respective locations in the presence of high G loadings. Pilots report blackouts in high-G turns (over 6 Gs) that are sustained for more than 25 seconds. Smokey Yunick conducted some research and concluded that the body can endure 50 Gs for 0.1 second and 100 Gs for 0.005 second. However, these are not blackout limits; they are the lethal limits. The organs suffer irreparable damage at these levels. It is imperative, then, that we find ways to reduce these forces in crash situations. (I don't know who Smokey's guinea pig was.)

Decelerative G forces in a crash can be reduced in two ways. The kinetic energy to be dissipated can be reduced or the length of the deceleration time can be increased. Both of these methods are commonly used.

Although road course wrecks often involve some runoff area before the impact, ovals and street circuits are usually lined with concrete and offer no runoff or cushioning whatsoever prior to impact. A 1700-pound race car traveling 150 MPH and encountering such a solid object releases over one million pounds of kinetic energy. To put it another way, if the car were to hit a concrete wall head on at that speed, the instantaneous load would be about 600 Gs. This is obviously not a survivable impact. When speaking of numbers this high, the deceleration rate of the car, or at least of the driver, must be as long as possible to reduce the peak and sustained G loadings on the body. Many purpose built race cars use "deformable structures," which are composite or monocoque structures, usually placed at the sides of the car or in the nose to crush and dissipate energy in a crash. NASCAR's "door bars" serve the same purpose. The theory is that, as these structures are deformed, the car is slowed at a rate low enough to prevent injury. And, by the time the structure is completely deformed and the chassis comes in contact with the intruding object, whether that object is a guard rail, a wall, or another car, the remaining G forces are not sufficient to cause injury. This is just theory, though. To lower the G levels to tolerable limits at racing speeds, these deformable structures would have to be *seven feet thick!*

The nose box on this Formula 2000 is designed to crush in the event of a frontal impact, dissipating some of the kinetic energy and reducing the possibility of injury. Notice also the shear plates attaching the suspension to the chassis. These will break in a crash, allowing the corner to leave the car and further reduce the energy to be dissipated.

In addition to having these crushable structures of the maximum practical size, a car should be designed so that the frame or aluminum/composite tub crushes to further dissipate energy. This is relatively easy to do with a steel space frame or an aluminum monocoque tub, although in the event of a crash, severe damage will usually result and be expensive to repair. This expense, however, is definitely preferable to the hospital costs and the pain that would be suffered by the driver as a result of injury. The downside of having the frame or tub absorb this energy is that it must still remain intact to avoid crushing the driver. This is an area where a great deal of thought, experience, and training in Mechanical Engineering and Finite Element Analysis is of considerable help in designing the structure to dissipate energy and still provide a "survival cell" around the driver. The surface has only been scratched in this area, and we will see rapid and far-reaching developments during the next few years.

Stand 21 makes some of the world's best safety equipment. They are very popular in Europe, and many pro drivers on this side of the pond use their equipment as well. However, I was quite alarmed recently to find that they are manufacturing seat belt assemblies that use fabric that has very limited stretch. This must be an attempt to prevent the belts from stretching to such an extent in a violent wreck that the driver's head and arms would be above the roll bar/cage. But it ignores the fact that one of the best methods we have of allowing the body to slow gradually on impact is to let the belts stretch. So, although most of their equipment is excellent, Stand 21 is missing the boat on these "no stretch" belts.

In addition to slowing the car/body gradually to prevent injury, we can also reduce the kinetic energy that must be dissipated in a crash. Indy cars in the '50s experienced some really spectacular and brutal flips, with cars catapulting into the air and cartwheeling down the track for hundreds of yards, before finally coming to rest. Each time the car returned to the ground and bounced back up again, the driver experienced massive G forces. When the car came to rest, it was usually completely intact, although severely bent. Drivers of that time often drove in tee shirts, used helmets little better than the day's football helmets, and had no fuel cells or fire extinguishing equipment. It's little wonder we lost so many drivers back then. Today's cars are designed so that the major mass assemblies, the engine, gearbox, and suspension, break away from the car, leaving only the driver's compartment to slow to a safe stop. Although a major part of the car, the driver's compartment usually only weighs around 30% of the total weight of the car, thus such a breakaway design significantly reduces the energy to be dissipated.

Kevin Cogan had a horrible crash at Indy in the late '80s. He hit the Turn Four outside wall, where the suspension came off. He then bounced into the inside wall, which separated the engine/gearbox assembly from the cockpit, then slid into the end of the pit wall at high speed, and finally came to rest against the inner pit wall on pit road! It was a spectacular crash, but Kevin climbed out of the bare tub of the car, took a few steps away, and looked back at the remains as if to say, "Did that really happen?" He survived uninjured, for the most part, and many feel that reducing the mass attached to the cockpit was largely responsible for his good condition.

These methods have proven to be highly effective in reducing injury in serious wrecks in the last few years. The best way to reduce the harmful deceleration forces, though, is not to design deformable structures into the car or to reduce the mass of the car in a wreck, but to build deformable structures around the objects that may be hit in a crash. Many club racing tracks use hay bales, water filled barrels, or tire barriers around immovable objects such as trees and phone poles. While these are better than nothing, a high-tech system is being developed for oval tracks which will eventually be adopted by promoters of street circuits and road courses. This involves replaceable sections of crushable walls made of a solid, expanded foam. The foam blocks act like giant airbags on impact, improving the driver's chances of survival, and reducing damage to the car as well. Several such systems are now being tested, including one at the new Texas Motor Speedway near Fort Worth, Texas, and at Indianapolis. When these systems prove their worth, and when it is economically feasible for the promoters (or when racers force the issue), many are sure it will become common. Until then, we must continue to try to reduce the forces acting on the driver in more conventional ways.

Will racing ever be safe? Absolutely not. It is safer now than it ever has been in the history of the sport, but drivers are still critically injured and killed each year. There is no way to foresee the myriad of ways in which a driver can be injured. Every accident is different. We can learn from our experiences, though, and explore the reasons for injuries as they occur. When trends are discovered, we can take precautions to reduce the possibility of a specific type of injury. For example, in a case such as the rash of head injuries in Formula One in 1994, due in part to the low cockpit sides of the F/1 cars, we can reduce the possibility of heady injury by raising the car sides. But raising the car sides will never eliminate these head injuries; it will just stack the deck against them. Racing will always be dangerous. We must continue to do all we can to improve the drivers' chances of survival in serious accidents. As long as man is moving faster than running speed, though, risk will be a part of racing, and that, of course, is part of its allure.

Accident Avoidance

There are two types of accidents in racing: single-car accidents in which the driver just loses control, and multiple-car incidents. Multiple-car accidents sometimes start with a single car and ultimately involve more. Let's look first at avoiding loss of control.

A few accidents are a result of mechanical failures, either direct results, such as when a suspension part breaks or a tire blows, or indirect results, as when oil is dropped on the track. The overwhelming majority, though, are caused by driver error. One way or the other, most accidents involve a loss of traction of at least one tire—either at the front end of the car or the rear. Accidents resulting from the driver losing traction at the front of the car can really only be caused by two things. If the car is understeering, regardless of the reason, the driver must change his line to compensate. Braking must be accomplished earlier and the turn in point should be earlier as well. If he does not change his line, a possible result is a collision with some object just outside the track out point. You will find more about lines in Chapter 6. The other cause is going into a turn with one or both front tires locked, which causes understeer and generates the same result. In this case, the driver should brake a little earlier and a little lighter to avoid locking a wheel.

Losing traction at the rear of the car is much more common. The first opportunity to lose the rear on approach to a turn is while braking. When it happens, a rear tire will lock (usually the inside rear) and unless the resulting oversteer is brought under control immediately, a spin will result. This is caused by the car being set up with too much rear brake bias, but the driver has direct control over how much braking force he uses. In this case, he should back off of the brakes a bit or, at least, be more sensitive, to avoid locking a wheel.

Another mistake that may result in a spin is not revving quite high enough when downshifting while approaching a turn. This momentarily slows the rear wheels, bringing the rear of the car to the outside of the turn. Since the downshift is made while the car is being slowed, it is likely that these two mistakes may combine. Too much braking force, coupled with a downshift made with insufficient revs, will almost certainly produce a spin. If it happens, it will do so very fast. Great sensitivity is required to correct it fast enough. If the driver did not have enough sensitivity to keep from locking a tire, it is unlikely that he will have enough to avoid a spin.

When the back end of the car begins to slide out, a novice driver's first reaction is to lift off of the throttle. This is precisely what he should not do. Study of vehicle dynamics tells us that when a tire is loaded with both accelerative and cornering forces, a sudden reduction in either will cause a reduction of the other. If the back of the car comes out, stay in the throttle. This may preserve enough traction to allow you to save it. It also means you cannot lift in the middle of a turn, because doing so will always cause an oversteering situation, which, if not corrected, will produce a spin. Obviously, this also means you cannot shift while cornering.

Other spins can be caused by using the wrong line through a turn and apexing too early. When a driver does this, he will run out of road at the exit and be forced to turn the car sharply to keep it on the track. Depending on chassis setup, this sharp steering input may also result in a spin.

If the car does run off the track at the exit, the outside wheel will have much less traction than the inside. To keep this situation from becoming worse, stay in the throttle and ease the car back on the track.

If you are driving through a turn on the limit, the rear of the car will be right on the edge of coming around. Many small corrections of steering angle must be made to keep the back end behind you. This takes nerve and sensitivity. A beginner has neither of these, and this is the reason track time is important early in a driver's career. Both of these qualities can be developed through practice.

In driving schools, instructors sometimes say, "When you spin, both feet in." As the car begins to come around, use the throttle to keep the engine running. Depress the clutch, since the rear wheels will soon be turning backward, and use the brake selectively to slow the car, when it is traveling either directly forward or backward. Being comfortable with the "toe and heel" technique is important. Provided there is sufficient room, you will roll to a stop, select first gear, and drive away. It is sometimes possible to "catch" a spin at either 180 or 360 degrees, but this usually requires that the car be still on the pavement. If you are in the grass or dirt, the scarcity of traction will prevent this. As a matter of fact, traction is so low in the grass that drivers sometimes joke about speeding up when they hit it. If this is the case, perhaps race tracks should be paved with grass.

If a sudden stop is imminent, it is wise to spin the car intentionally to hit with the rear of the car. In a production-based car, there is usually more crushable material to the rear, and in a purpose built competition car this helps to save the driver's feet and legs. In both cases, the loads can better be distributed into the driver's torso with the back of the seat than with the belts.

A sudden stop does not include a car spinning in front of you. In this situation, you must consider the other driver's safety as well as your own. Many times, you will be required to either slow the car so rapidly or change line so drastically that your own car will spin to keep from hitting his. If you can see no choice but to hit the other car, try to do it so as to endanger the other driver the least.

Many racers seem to be afraid of open-wheel race cars. This is probably because they do not have a proper knowledge of how to run close. Formula Car drivers typically have an attitude that allows them to run within an inch of the other car as long as they do not touch it. The reason for this healthy respect is the danger of tires touching front to back and launching one of the cars. No one has ever benefited from taking such a ride. It does not happen often, fortunately, and their

Reflections on Safety

In the early part of the '95 season, the motorsports community reeled from two similar accidents, each of which brings up serious concerns about driver safety. At the inaugural Formula Atlantic race of the '95 season at Miami, an experienced driver, Bob Ferstl, spun and hit the inside wall of the street circuit, lightly damaging his car. He pulled away, apparently trying to make it back to the pit lane, and stalled. Another car came through the turn at high speed and hit Ferstl square amidships, severely injuring him. Although reliable information is scarce, it seems that the yellow flag may not have been displayed after Ferstl moved away from the wall.

Barely six weeks later at the IMSA race at Road Atlanta, two GTS-2 cars tangled in Turn Twelve at the beginning of the pit straight. Fabrizio Barbaza in a Ferrari WSC car spun to the inside to avoid the incident. Jeremy Dale in a Spice-Olds WSC came down the hill toward Twelve at high speed, unaware of the spinning cars ahead of him. Entering the turn, he saw the GTS-2 cars on his left, but was unaware of Barbaza in a cloud of tire smoke on his right. Dale veered to the right away from the wreckage and bouncing tires, and the Ferrari continued to spin out of its tire smoke and appeared in side view directly in front of Dale. He was approaching at about 140 to 150 MPH and had barely more than a car length to slow before impacting the Ferrari dead center. The impact was so great that Dale's car was completely destroyed as far back as the instrument panel and Barbaza's car was literally cut in half at the roll bar. Both drivers suffered severe injuries.

We can learn several lessons from these devastating accidents. Ferstl would have had much less chance of being hit had he stayed against the inner wall of the turn until the corner workers arrived. This would have ensured that the yellow flag would have been out. Unless visibility is unlimited or you are in a very dangerous place, do not move your car after a spin or an accident until instructed to do so. This is also a good time to point out how dangerous it is to try to get out of a stalled or damaged car on the track. *NEVER* release your seat belts unless off the racing surface or instructed to do so by a corner worker. They know when it is safe to get out of the car. Think of the grim possibilities if Ferstl had been partially out of the car when it was hit.

At Atlanta, the accident happened much too quickly for flags to have helped. Dale's first clue that trouble was brewing was that the cars were throwing debris onto the outside of the track. His attention was focused on them and he failed to see the significance of the cloud of tire smoke at the inside of the turn or the set of spiraling black marks on the pavement. He thought he was past the danger point and opened the throttle again before seeing the spinning Ferrari. Had he paid more attention to these signals, the impact might have at least been less severe. When approaching a crash scene, it may be difficult to know where all the dangers lie. Unless corner workers are scurrying around the cars, *SLOW DOWN* until you are certain that you are completely past the scene. If corner workers are present, of course, you will be forced to slow down and will know without a doubt where the accident is.

It is accepted as fact that yellow cars are the most easily seen, but careful analysis of the video of the Atlanta accident may show that other colors might be preferable in certain circumstances. Barbaza's bright yellow Ferrari had little contrast with its own tire smoke. Had the car been red or possibly even a multicolor scheme it might have been more easily seen in this case.

As has been mentioned, modern racing cars are designed so that the kinetic energy to be dissipated in a crash is reduced by reducing the car's mass. A very effective way to accomplish this is by allowing the engine/gearbox/rear suspension assembly to detach itself from the driver's compartment. When Barbaza's Ferrari was split in half, it did exactly what it was supposed to do. Of more concern with the present generation of race cars, however, is the integrity of the front end, which is intended to protect the driver's feet and legs and the side impact protection, which protects the drivers torso. Stan Fox's horrible crash at the '95 Indy 500 also demolished the front of his car back past the instrument panel, leaving his legs bouncing along the pavement and smashing into the outside wall. Although he did sustain severe head injuries due to the high g-forces that destroyed the front of the car, it is a miracle—one that nobody seems to be able to explain—that he survived this accident without severe injuries to his legs. While no car can be expected to withstand a head-on collision with a concrete wall, or even another car, at 150 MPH, perhaps we can and should build cars that better protect the driver's forward extremities in these situations. A few years ago, most sanctioning bodies passed rules defining the forwardmost allowable point of the driver's feet. This is usually the front of the front wheel rim or the front wheel center line. Many years ago regulations were instituted detailing the construction of "deformable structures" to be mounted at the sides of Formula Atlantic cars.

These specifications are certainly a step in the right direction, but it seems we need some minimum crashworthiness specs as are used in Formula One to ensure that these devices are doing their jobs. Each Grand Prix team is required to present a bare carbon fiber tub to FOCA, the governing body of Formula One, for destructive testing. At the Cranfield Impact Centre in the UK, the tub is attached to a moving platform, and is subjected to a series of test collisions to determine how well it protects the occupant. The tub must exceed FOCA's requirements before the team is allowed to compete with a car built around the type of tub being tested. Many Indy Car and other type of chassis constructors do load testing with finite element analysis computer technology and can predict the severity of chassis damage at a given load. If we used standards and real world tests similar to those used in F1 for other classes, though, perhaps Jeremy Dale would not have sustained multiple fractures of both legs and Stan Fox would have been protected instead of just lucky.

respect for each other's tires is the reason. Just stay out of a position where one of you could get launched and you will not have a problem. Banging wheels rim to rim does nothing except bend suspension.

Insurance

The interaction between health care providers and insurance companies has changed a great deal in the last few years. HMOs and comprehensive insurance programs are now the order of the day, but the metamorphosis that brought these about may not yet be complete. All racers should look very carefully at their insurance coverage because of this changing climate. Although most major sanctioning bodies cover competitors, their coverage is mostly limited to supplementation of an individual racer's personal coverage.

Many health care policies contain exclusion clauses whereby claims involving hazardous activities are not covered. These can include sport aviation, sky diving, rock climbing, scuba diving, and, of course, auto racing. Some insurance companies have no such exclusion clauses or exclude one sport, but not another. Check your policy carefully to ensure that claims arising from auto racing will be covered. It is best to buy insurance that covers the sport prior to beginning a racing career. All too often, racers do not check their coverage until after they begin to race, and then find, when it is too late, that auto racing is excluded.

With health care costs spiraling upward at an alarming rate, it is important to have adequate coverage. These days, the treatment of a serious injury may cost $500,000. Just having a major medical policy that has a high dollar ceiling may not be totally adequate, though. Some of these have high deductibles in order to keep the premiums low, and these deductibles may take many forms. Some policies only cover hospital costs from $5,000 or $10,000 and up, and have other deductibles for medication. Other policies may not include doctors' bills or may have special provisions for surgical procedures. For the layman, these special clauses and provisions are difficult to master. It is best to have an insurance agent you know and trust design a policy for you that is suited to your particular circumstances.

Those circumstances include the coverage afforded by the organization with whom you intend to race. Most small organizations, such as local oval tracks, only have liability coverage under which spectators or crewmembers are covered. Rarely are competitors covered in these events. Larger organizations, such as SCCA and USAC, provide insurance for the competitor as well, but only in addition to the individual's coverage. USAC provides $25,000 excess coverage for an on-track accident, and PRS has a similar arrangement. SCCA is at the top of the heap in terms of benefits, with $5,000 paid in addition to other insurance and excess coverage of an additional $45,000. Another SCCA policy provides excess coverage up to $500,000 and a death benefit of $25,000 is also included. When joining an organization as a driver, ask to see the specifics of its insurance coverage in order to help your agent design your own policy.

Part Two: The Four Things That Every Successful Driver Learns

Chapter 6

Learning to Drive

Use the proper line around the track, not just through each individual turn, maximize traction in all three phases of the turn and regulate speed into the turn with the brakes just enough to be able to change direction with as much speed as possible.

Driver's Schools

Training has become very important in our culture. We get training for everything from business administration to cake decorating. Until the last couple of decades, though, race drivers were self-taught. This is not the case today. Through the efforts of Jim Russell in England in the '60s, racing driver's schools have now taken their place with all other academic institutions. Several drivers's schools around the country offer programs ranging from a single-day orientation course to a five-day advanced session.

Racing is such a complex sport that some have proposed that a student study and practice for four years to receive a baccalaureate degree, at which point he would be awarded a Pro License and be allowed to race for a living. This proposal is really not too far-fetched. The Skip Barber School has formed its own racing series, the Barber Dodge series. After successfully completing the basic school and the advanced sessions, students are allowed to enter a series of races for a full season. The cars are prepared by the school and are as equal as possible. As with any form of intensive education, these racing series command a high price. Be prepared to spend your college fund on racing, if you decide to take this route.

If your budget does not allow you to enroll in one of the school's racing series, the two-, three- or four-day courses are a good place to start. In them, you will be taught how to make the car do what you want, and you will also learn why to do it. You should be aware, however, that the schools are as unique as their instructors. Each school has its own curriculum and teaches its own methods for car control, and not all of them are the same. Before deciding on a school, investigate what they teach, what type of cars they use, the record of the instructors, and how long they have been in business. Talk to drivers who have trained with them. When the schools tell you that Michael or Al, Jr. studied with them, remember that they attended *all* of the schools, an important consideration is if you can afford it.

Driving instruction today is intensive and effective. Many driving schools offer exceptional programs for those aspiring to become pros as well as for the serious amateur.

Many schools use Formula Fords or reasonable facsimiles as their primary training vehicles for the regular courses. If the school cars were configured as race cars are, the tire and engine bills would severely limit their ability to stay in business. For this reason, street tires and low RPM limits are used. When you get into the same type of car in a real race, both lateral and longitudinal acceleration, as well as speeds, will be noticeably higher.

The Russell and Barber schools have each produced videos that describe what happens in their beginner's courses. Both are informative and include some good in-car footage. They should be on every racer's video shelf along with *Grand Prix* and *Le Mans*. See Appendix A for a list of schools and Appendix B for a recommended reading list and details on the videos.

Car Control

It may seem too basic, but a new race driver must learn to use the car's controls properly. This is not to say that drivers of street cars use the controls improperly, but just as a race car is very different from a street car, so are the conditions and the objectives on the track. A driver new to competition has some things to learn.

Steering

Driving smoothly is of paramount importance to achieving quick lap times, and the reason for this is found in basic physics. Millikens' *Race Car Vehicle Dynamics* is now in use at several colleges and university as an undergraduate textbook in vehicle dynamics. In it, you will learn that traction and therefore cornering force that any particular tire is able to generate is directly proportional to the weight supported by that tire. Actually, a graph of that relationship would be a curve, with the shape of the curve being controlled by the manner in which the car has been set up and what the tire engineers designed into the tire. (More about this in Chapter 8.) The main thing to remember now is that tires don't like abrupt weight changes, and show it by a reduction of traction. These weight changes are *always* caused by the driver's control inputs. If he is jerky and abrupt with the steering, the tires respond with a reduction of traction, just when it is needed most.

The steering ratio of a competition car is usually substantially quicker than that of a street car. This means that the front wheels turn a greater angle with the same movement of the steering wheel. A quick ratio makes the car more responsive to steering inputs and accentuates any abrupt movements of the steering wheel. A popular adage goes like this: "To be fast on the track, you must be slow in the cockpit." Plan your movements with the wheel and make them methodical and gentle. It is very difficult to be gentle when you are in an environment in which you are pushed against the belts and other parts of the driver's compartment with enough force to leave bruises, but it is also essential.

The function of the steering wheel is to control the direction of the car. The function is not to give the driver something to hang on to in order to keep himself in position. That job belongs to the seat and belts. On some parts on the track, the dead pedal can be used to help support your body, but you should never brace yourself with your arms. If you are pushing on the wheel to hold your upper body in place, your steering inputs will tend to be jerky and imprecise. Make sure the seat is positioned correctly, and the belts are tightened properly to leave your arms free to control the wheel and shifter.

Throttle Control

One major goal of racing is to get the car around the track as fast as possible. To accomplish this, the driver must stay on the throttle as much of the lap as possible. Not just burbling part throttle, but on it hard. In reality, many things interfere with accomplishing this task, one of which is upsetting the rear tires with gobs of torque when they are near their traction limit. For this reason, the throttle cannot be slammed open, but instead, the car must be given only as much throttle as it can use at that moment. The more powerful the engine, the more difficult this task is. Open the throttle as much as is possible without generating oversteer. If the car oversteers too easily or if the inside rear spins, remember exactly where it happened and what you did before the onset of the problem and tell your engineer. Every patch of pavement on the track has its own traction characteristics, and a car that is on the verge of oversteer will step over the limit many times through a turn exit, requiring the driver to make constant small changes to steering to maintain the chosen line. If the car feels somewhat unstable and barely in contact with the pavement, you are doing it just right.

Braking

Carroll Smith says that the last thing a driver learns to do really well is brake. Many drivers have a problem with shifting, too, but braking is certainly one of the more difficult and complex tasks required to go fast. This is an area where street driving offers little help in preparing a driver for what happens on the track. In order to be smooth, the driver cannot slam on the brakes any more than he can the throttle. Braking causes weight to be transferred to the front of the car, unloading the rear tires. This is the reason the braking force must be biased toward the front wheels. Even though the front tires are usually smaller than the rears, they do most of the work of slowing the car because most of the car's weight is on them under hard braking. With a large amount of weight on the front tires and very little on the rears, the car is quite unstable. It can be controlled much better if the weight transfer happens smoothly. To begin braking, ease down on the pedal, but do it firmly and quickly. When the nose of the car has dropped to its minimum level, the driver should be on the pedal *hard!*

Brake as hard as possible without locking a wheel. This is like saying you should tighten a bolt until just before it breaks. Just as a bolt will give some warning signs that the end is near, though, so will a tire. When a front tire locks while braking in a straight line, the uneven braking forces will cause the car to pull to one side, which will be felt in the steering. If any cornering whatsoever is being done, weight transfer will cause the inside tire to be the problem, and this can be felt in the steering, through the understeer that will inevitably result. Both conditions can be felt in their beginning stages before the tire has come to a complete stop. Up to the point of lock up, though, the weight transferred to the front wheels will actually improve steering response. (More about this in Chapter 8.) When a rear tire locks, it leads to oversteer without much warning, regardless of whether the car is being slowed in a straight line or during cornering. After a wheel locks, the car will not decelerate as quickly, and you may have little control over the direction of travel. Release the brakes slightly to let the wheel(s) roll again.

A somewhat inexperienced driver once asked, "How do you know how much to brake for a corner?" Finding the answer to that question is not easy. Braking can more accurately be called regulating speed. The object is to reduce speed on entrance to the corner just enough that the necessary change of direction can take place. The speed that remains after the brakes are released should be the car's minimum speed in the turn.

The throttle should be kept open for as long as possible upon approach to a corner, and braking should begin immediately after lifting off the throttle. The point on the track at which that takes place should be moved closer to the corner on successive laps until one of two things happen. The most probable is that oversteer will cause the speed or the line through the corner to be adjusted to maintain proper attitude of the car. In this instance, the proper amount of braking has been determined, which produces the maximum speed possible at the point of direction change. The point of brake application should be just slightly earlier so that the car is on the verge of being out of shape, not over it.

The other possibility is that braking may be started too late and continued too deep into the corner in an effort to bring the speed down sufficiently to make the corner. If this happens, the proper line cannot be maintained. This time, braking should begin (and end) substantially earlier. Going into a corner with the car crossed up and off the proper line leaves room to pass on the inside and is hopelessly slow. It is always better to brake a little too early, rather than too late. Some drivers, though, brake much too early and too lightly, and are regularly passed on entry to turns. If the car is not at its limit through the turn, you are braking too early.

Unless you are braking too lightly, you cannot make significant improvements by going into a turn deeper or using the brakes harder. Improvements are made by improving cornering speed and using the brakes to adjust the car's speed to a new higher cornering speed. Aryton Senna explained finding the limit this way, "When you feel like the road is escaping you, when you say to yourself, 'That's it. I'm going off.' but then instead you stay on the track, that's when you've reached your limit; you've gone as fast as you could."

Trail Braking

Most of the braking for any given turn will be done in a straight line. When it is *all* done in a straight line, it is called the European driving method. This is the proper braking technique for a very slippery track, as when it is wet. In the dry, however, all top drivers instinctively use *trail braking.* Although this term has become a buzzword in the last few years, it is neither controversial nor is it complicated. Trail braking is simply the technique of beginning to turn while the car is still being slowed. As the steering wheel is turned in toward the corner, the driver eases off the brakes. For a short period of time both the brakes and steering are being used.

To understand trail braking we must use a concept that Mark Donohue originated in the late '60s—the traction circle. If we plot the traction that the tire/road system provides in all directions on a Cartesian graph, we get Figure 6-1. In this case, we have 1.25 Gs of traction available in any direction. The driver can use all of that traction in braking or all of it in cornering. Using 1.25 Gs for both at once, however, means the tire/road would need to generate about 1.75 Gs. Since this is beyond the capabilities of the system, a locked wheel and a possible off-course excursion will be the result. In this example, to stay at the limit of adhesion into the corner, the driver must ease off the brakes as the steering wheel is turned to use only the 1.25 Gs available. This keeps the car on the circumference of the traction circle, which is the limit of available traction. Most drivers do this instinctively. The better the driver is able to keep the tires at or near the circle during combined braking, cornering, and acceleration, the lower will be his lap times.

The exceptions to this technique are when approaching a tight turn where it is impossible to brake into the corner and when the track is wet or has been oiled down. Occasionally, a top driver in his class says that he never trail brakes. In most cases data acquisition will disprove his claim. In the others, it is likely that he drives a car that is not known for having either good brakes or particularly good handling, making trail braking difficult. Formula Vee is the best example of such a car.

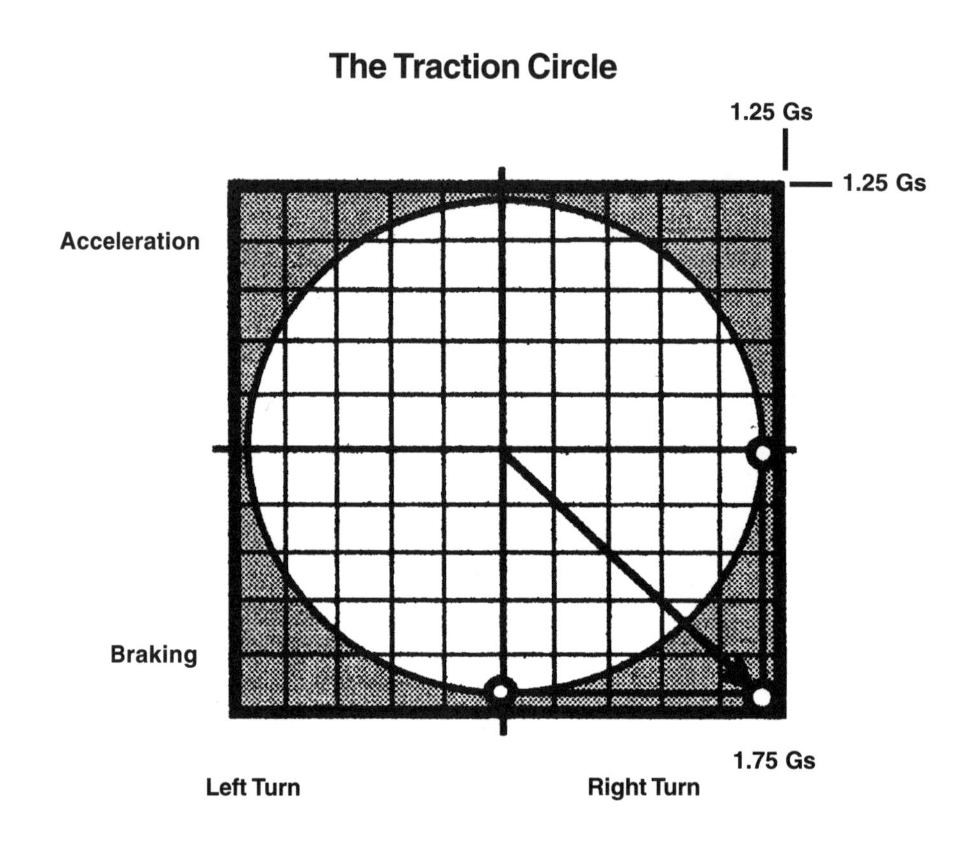

Fig. 6-1 Combining any turning at all with maximum braking results in operating outside the traction circle. This leads to locking a tire and possibly worse.

Shifting

All transmissions, called "gearboxes" in racing lingo, whether built specifically for racing or derived from street transmissions, are of the constant mesh variety. This means that all of the gear pairs are always meshed with each other. The gears on one shaft are coupled to it by splines, and the others run on bearings on their shaft. Shifting takes place by moving a splined hub, which is coupled to its shaft, into face contact with one of the free spinning gears. Face dogs on both the hub and the gear engage to provide power flow. A racing gearbox is totally non-synchronized. It provides very crisp, positive shifts, but would be too abrupt for use on the street. Synchronized street gearboxes are sometimes used in race cars, and the basic difference in shifting mechanism is the synchronizer ring used between the hub and gear. Syncro rings make leisurely shifts smoother, but also tend to slow down quick shifts. A synchronized box is easily detected by the slight double click of the shift lever as a gear is engaged. Regardless of what type of gearbox is used, a basic knowledge of how it works is very beneficial. Once you realize how precise and delicate the shifting mechanism is, you will be less likely to use the "tear it out by the roots" technique.

Upshifts in a race car are very easy and much like those performed in a street car. The only real difference is the requirement of watching the tach so that the shift can be made at the right time. As you gain experience in a given type of car, you will know by the note of the engine when it is about time to shift. A quick glance at the tach will verify that the engine is a hundred RPM or two below the red line. Avoid the temptation to powershift a race car. Reserve that particular technique for your '57 Chevy.

The clutch operating mechanism should include a positive stop to avoid overstressing the pressure plate springs, but it should be adjusted so that the clutch can be fully disengaged. If your car uses a competition clutch, rather than a street or hot rod variety, you should remember that it is an on-off switch and is not intended to be slipped leaving the pit area. Slip the tires instead, but that is not a license to do smoky burnouts leaving the pits. When making a shift, it is not necessary to depress the clutch pedal to the stop. A simple tap to unload the mated gears will do it.

Downshifts are a little more difficult than upshifts, especially with a non-synchronized box. First, we need to dispel the myth that a driver uses the engine compression, by downshifting, to slow the car. Engine braking was necessary in the '50s when cars were heavy and used drum brakes. With today's cars, the brakes slow the car. The downshift is made in order to be in the correct gear when the power is applied.

To properly make a downshift, the engine RPM must be matched to the road speed in the lower gear. To do this, the throttle needs to be "blipped" while in neutral between gears. This is where the "heel and toe" technique comes in. It takes some fancy footwork to manipulate two pedals with one foot. The ball of your right foot is used on the brake, and the right side of your foot rolls over on the throttle pedal to bring up the RPM. (Thus, "heel and toe" is really an incorrect term.)

If the brake system is set up to use a large amount of pedal travel, it is difficult to judge where the throttle pedal will be as the car decelerates. A heel stop can be used to provide a reference point. Many drivers prefer to set up the hydraulic ratio of the brake system so that the pedal travel is minimized to avoid this situation. A side effect of this setup is that it takes a lot of force on the pedal to slow the car. When the pedal is very firm, the seat should be quite stiff to avoid pushing yourself back into it when you are on the pedal hard. All of this may seem like a lot of work just to make the "blips" more accurate, but it is important. If you cannot match the RPM properly, the car will jerk as the downshifts are made, upsetting the balance of the car. If you over blip, assuming it actually goes into gear, the car will lurch forward as you are trying to slow. If you under blip, the rear tires will slow momentarily, lose traction, and possibly cause a spin.

Some drivers like to hit each gear in succession when downshifting. Others will skip gears. Most cars that use production based engines are typically shifted by skipping gears and using one big blip. It seems logical that the peaky four-valve racing engines, with small flywheels and clutches, rev so quickly that each gear should be selected when shifting down. An informal survey of Formula One drivers before sequential gearboxes (that means standing at a corner and listening) indicates a pretty even split between those who do and those who do not. That is changing in these days of the sequential shift, though. In most cars, medium speed corners require such short braking distances that the driver does not have time to make more than one downshift. When encountering a corner where it is necessary to go from fourth to first, many drivers go from fourth to second to first. This keeps the RPM in the power band and allows the driver to make two nominal sized blips instead of one gigantic one. If your car uses a racing gearbox in which the ratios can be changed, remember that the required RPM to match road speed changes when ratios are changed.

Ask any ten drivers how (or whether) they use the clutch when shifting, and you are likely to get about a dozen answers. Skip Barber recommends double clutching and teaches it in his school. Craig Taylor is a former National Champion in Formula Continental and has also won the USAC F/2000 West Championship. Being in the gearbox business qualifies him as an expert witness on this question, and he would cheat on his wife before he made a shift without the clutch. (And Craig is *very* happily married!) Price Cobb has been a paid professional driver for almost a decade and has driven GTP cars in the U.S. and won Le Mans. He knows how to go fast and how to make a car last. He says that when nursing a car or trying to make it last 24 hours, he uses the clutch on every shift. If he is not as concerned with the longevity of the dog rings in the gearbox, he uses the clutch on downshifts to give himself a little room for error on matching RPM, but does not use it on upshifts. When qualifying or really putting on a burst of speed, he does not use it going up or down. Jacques Villeneuve, says, “I don’t ever use the clutch for gearshifts because I left foot brake.”

In reality, you should probably use whatever method makes you feel the most comfortable. My personal opinion is most in line with Price’s; what I do with the clutch depends on the circumstances. Dave Salls always used the clutch in his dog engagement gearbox, although he felt he might have been slightly faster without it. Changing gears (and dog rings) in his car required pulling the engine, and that was too much trouble for the slight benefit in lap times. If you are using a synchronized box, it is possible to shift it without the clutch, but I would not recommend it. I used to practice shifting my street car without the clutch, just so I would remember how to do it with synchros. When I bought a more expensive car, though, I stopped.

This entire discussion of shifting assumes that the RPMs are always properly matched. If they are not, you will tear up gears, dog rings or synchro rings. Learn to match RPM—or learn to double clutch.

Alan Johnson's Turns

For decades, race drivers took each turn on the race track as an individual exercise. For any given turn, there is an ideal, theoretical line which maximizes speed through it. The idea is to use the longest turning radius possible, so that the car does not have to be slowed any more than necessary. In the late '60s, Alan Johnson ran a 914/6 Porsche in, what was at the time, C Production. He won four National Championships. In his book, *Driving in Competition,* he describes another cornering method besides the classic, ideal line and prioritizes three different types of turns in terms of their importance to lap times.

To begin with, a driver should think of every turn as consisting of three phases: the entry, the mid-part, and the exit. Dividing a corner in this way will be valuable, not only in terms of driving technique, but also in determining what changes should be made to improve the handling of the car. (More about this, too, in Chapter 8.) The important points marking these three phases of the turn are the turn in point, the apex, and the track out point.

Mr. Johnson told us that we should look at the corners and straights on a track together to minimize lap times. Assuming there is straight after the corner in question, if the speed at the exit of the corner can be maximized, the speed all the way down the straight is higher. A couple of MPH at the turn exit may turn into six or eight at the end of the straight. In order to accomplish this increase in exit speed, the apex, or point of minimum radius, must be made later in the middle part of the corner. In tight corners, the apex will only be a point. In longer radius corners, it will be an apex area, sometimes several feet in length. By driving deeper into the turn before the turn in point, a later apex can be used, and a larger radius can be used on the exit. Although the radius at the entrance of the turn must be tighter and the speed reduced in that area, the larger exit radius allows the throttle to be opened earlier, which produces the higher exit and straight away speeds. A late apex is used in a turn leading on to a straight. Johnson calls this type of turn a Type One. This technique is shown in Figure 6-2.

Conversely, a turn at the end of a straight requires that an early apex be used in order to keep the speed up as long as possible into the turn. An early apex sacrifices speed at the exit of the turn and should only be used when a turn at the end of a straight is followed closely by another turn. Johnson calls this type of turn, in which an early apex is used, a Type Two turn. If the turn is in between two straights and can be driven as either a Type One or Type Two, it should always be treated as a Type One. It should be apparent by now that, since the straights are where speed is high, and most of a lap is made up of straights, the lines through the turns are manipulated in order to maximize the speed where it counts. See Figure 6-3.

A Type Three turn, Johnson tells us, is a turn that occurs between two other turns. In this case, the line through the Type Three turn should be sacrificed to use the proper line through the Type Two that inevitably precedes it or the Type One that follows. Since far more time is spent on straights than in Type Three turns, lap times will be reduced in this way. In some cases, more than one Type Three turn will lie between the two more important ones. In such cases, the line chosen should

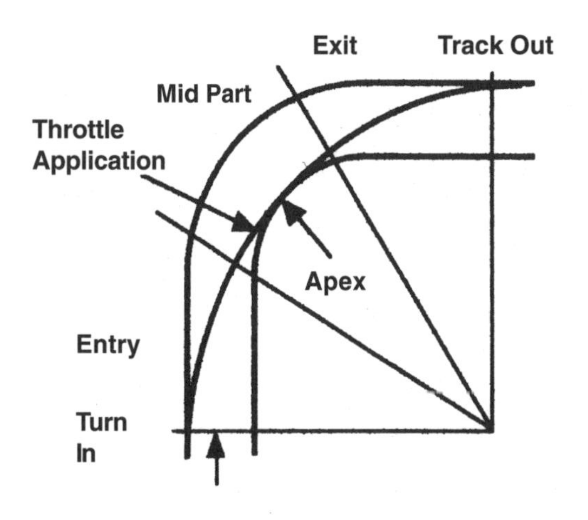

Fig. 6-2 (left) The ideal theoretical line maximizes speed through the turn.

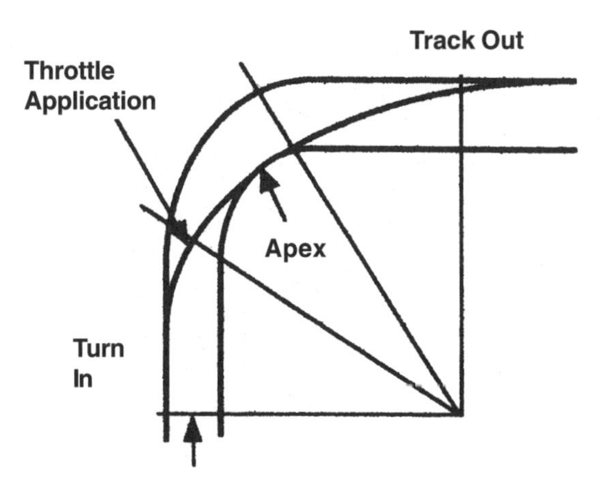

Fig. 6-2 (right) A Type One turn requires a late apex so the driver can apply throttle sooner to increase speed at the exit.

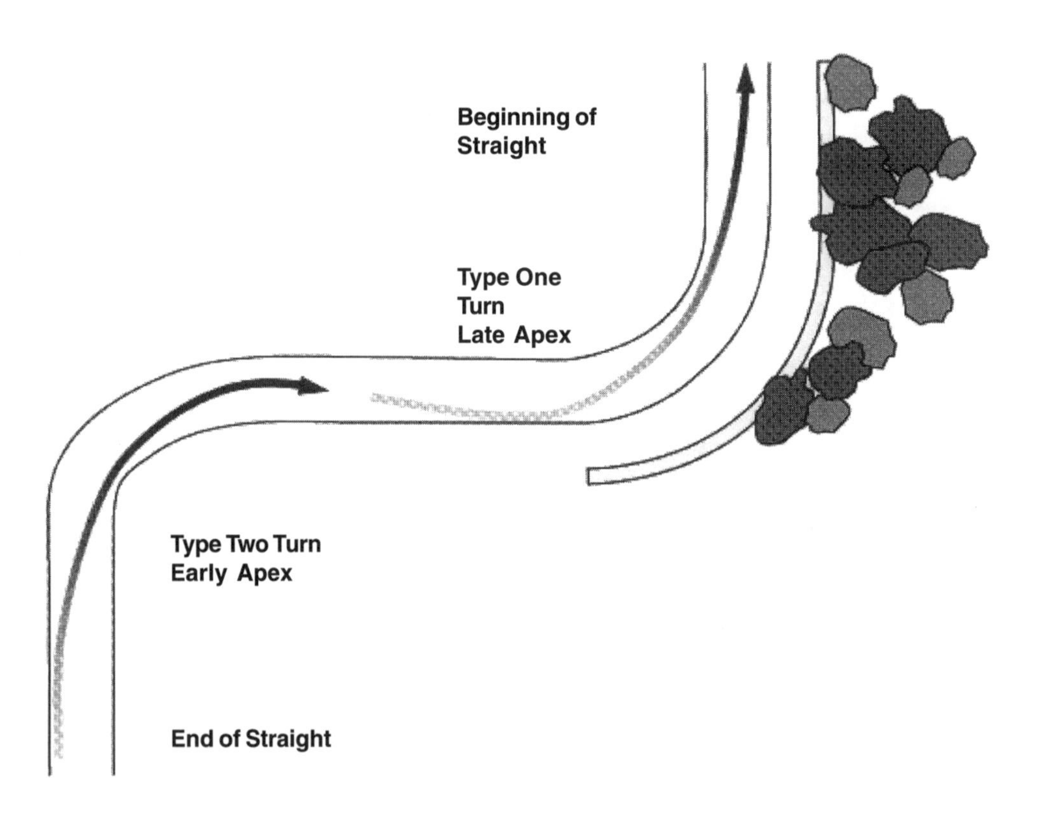

Fig. 6-3 A Type One turn always takes precedence over a Type Two. If a turn occurs between two straights, it is treated as a Type One.

always make the best use of speed onto and off of the straights. However, just because these turns are less important to lap times than the Type Ones and Twos, does not mean they have no effect on the times at all. Don't make the mistake of thinking you can relax through a Type Three turn and throw it away. Time lost is just that: *lost*. If three Type Threes were ever to be found together, the one in the middle would be taken using the classic line to maximize speed through it.

Figure 6-4 shows a Type Three turn between a Type One and a Type Two. It is interesting that the proper line (in this case) is almost identical regardless of the direction of travel. When the direction of travel is reversed, the Type One becomes a Type Two, and vice versa, but the Type Three is, of course, still a turn between turns. The track in Hallet, Oklahoma is run in either direction, as are many others across the country. It is important to realize that some alterations to line are required when the direction of travel is reversed, but the turns at the beginnings and ends of straights are taken with much the same line, just in the opposite direction.

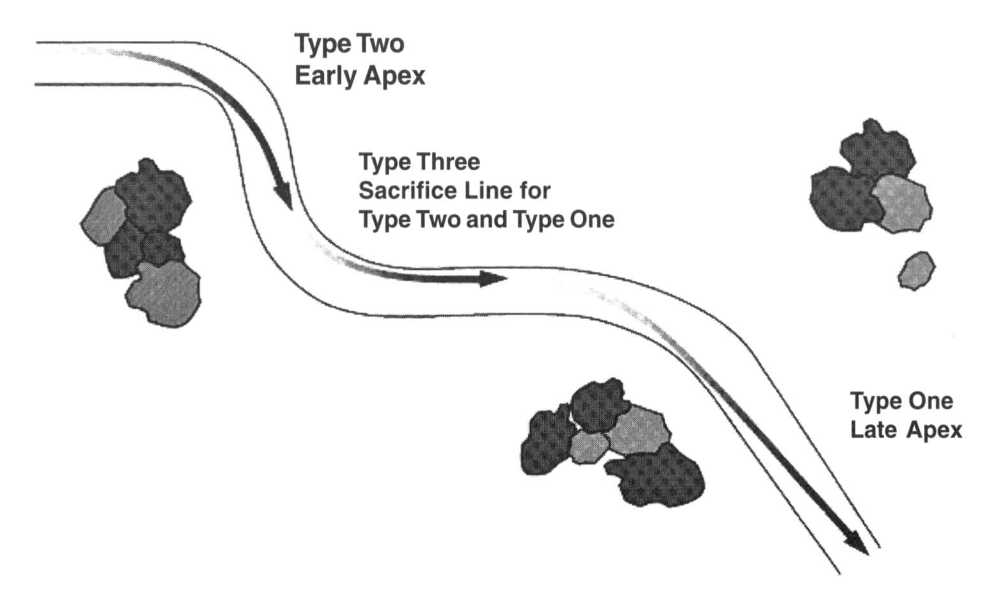

Fig. 6-4 When a turn is found between two others, the first and third turn always take precedence.

The Real World

If all a driver had to do was prioritize the turns on a track to determine the proper lines, racing would be relatively easy. Each of the turns we have just examined is a little different in the real world, however. Turns are complicated by elevation change, roughness, traction, bumps, dips, hills, track camber changes, speed, and the car's handling characteristics. The driver must sometimes compromise the proper lines in order to cope with these situations, and he must do so while accelerating, braking, and cornering at high force levels.

The turn depicted in Figure 6-5, for example, represents a turn at Lime Rock Park in Connecticut called the Downhill. In most cars, it will be approached at the top of third gear or the bottom of fourth flat out, with no braking. As the name implies, this turn goes downhill toward the entry to the main straight, so the car accelerates more rapidly than would be expected in these gears. It's a fast turn. On many tracks, the straight is just a bit wider than the turns, or an extra patch of asphalt has been laid down at the track out point to give the driver a little extra cushion. The Downhill is not built this way. If anything, it seems to tighten up a bit at the exit, just at it flattens out, which means the track out point must be a little closer to the inside of the turn than it looks on approach. It's also a bit rough. A wooded area lines the outside of the track behind a fence just a few feet from the track. These conditions make it a very tricky turn. The first few times through, a driver is tempted to lift before the apex. Due to aerodynamic drag, the car will slow rapidly in upper gears when the throttle foot is lifted, which is detrimental to speed down the straight. Another tendency is to apex too early and run out of road at the track out point.

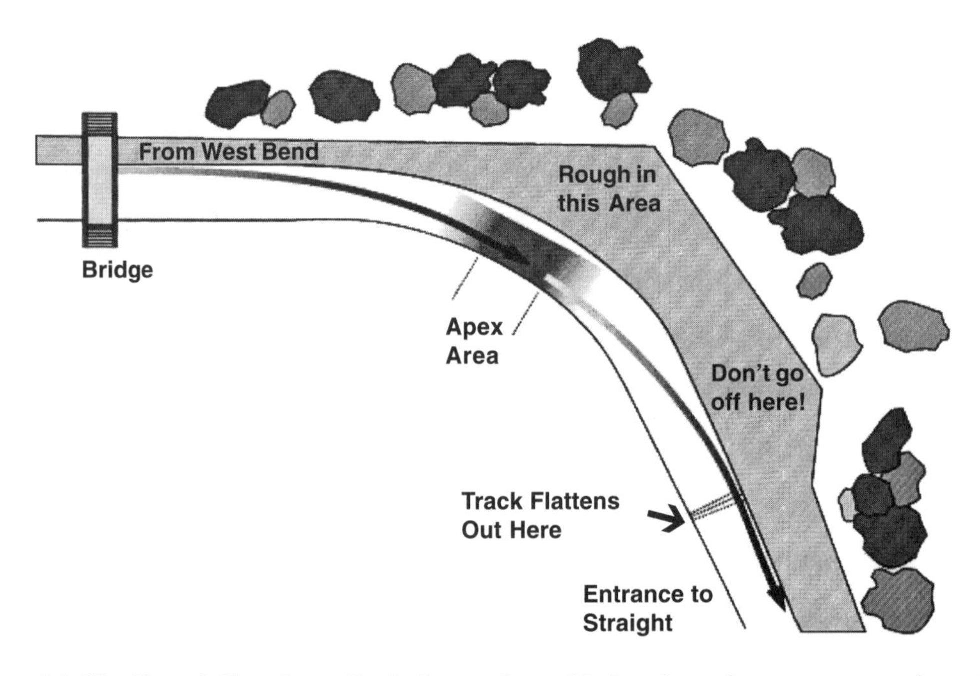

Fig. 6-5 The Downhill at Lime Rock drops about 20 feet from the entrance to the exit. In most cars the turn is approached at the top of third or bottom of fourth gear. The speed, elevation change, and roughness make it a difficult Type One turn.

The real trick to doing the Downhill right, after overcoming the temptation to lift, is to avoid scrubbing off any speed. Since this turn is taken quite fast, any scrubbing of the tires on either end of the car will slow it through the turn and down the following straight. Since the Downhill is taken flat out, the West Bend that precedes it can be considered a Type One turn also. The faster the car or the less downforce it creates, the more important the Downhill is to lap times. To be fast at Lime Rock, a driver must place the car with precision through West Bend and the Downhill, and be gentle with the car to avoid upsetting it.

In slower cars, Lime Rock Park's Downhill is not too difficult. When approached at high speed, though, a lot happens in a short span of time, requiring planning, practice, and precision driving for a good exit speed. Mistakes here have been disastrous for some drivers.

The Type Two turn shown in Figure 6-6 is Turn One at Road Atlanta. It is particularly difficult because it is approached at high speed in fourth or fifth gear. A driver must begin braking into it approximately at the pit exit just before the road drops into the "bowl" of the turn. He must turn in before it really feels proper and make a very early apex. Turning in this early feels uncomfortable because the exit of the turn is obscured at the turn in point by the bridge. The apex is just prior to the bridge abutment, and you must be only a few feet away from it. At this point, the track begins to climb steeply, and this banking effect helps contribute to the high speeds through the corner. At the track out point, the car must be turned back to the center of the track and a lower gear must be selected to be able to pull the hill and set up for Turn Two. Many drivers have gone off the outside of One, which is particularly dangerous because of the trees at the bottom of the hill outside the turn. In 1990 I worked with a Formula Continental and clocked the car with radar at the apex of Turn One at 108 MPH! Formula Atlantic cars, with their generous downforce, go through even faster in fifth gear. It is one of the toughest turns found among North American tracks, but looking at it in the diagram does not show its true nature.

To be fast, a driver must choose the proper lines through the turns and combine them into a line around the track, and he must also be able to deal with all of these extra conditions which slightly revise the proper theoretical line into a workable, practical one. Some drivers' mental images of the proper line is a white stripe four inches wide through a corner, as left by a paint brush behind a truck. They feel that as long as they keep the center of the front of the car somewhere near that imaginary stripe, they are on the proper line. For faster drivers that image is a stripe exactly as wide as the car. If any of the four tires get outside of that stripe, they are losing speed. A driver

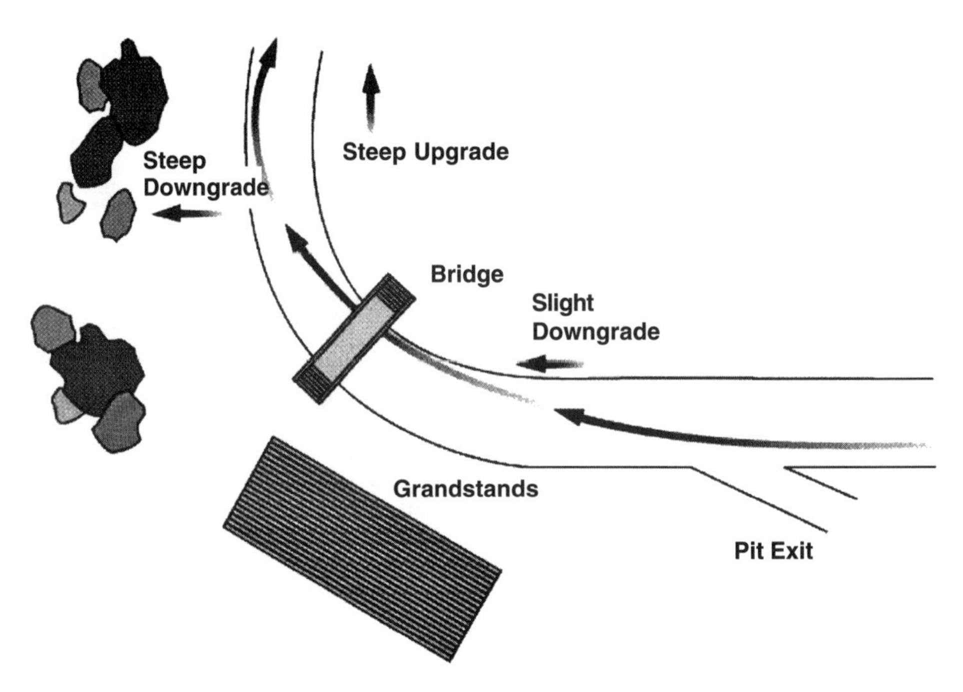

Fig. 6-6 Road Atlanta's Turn One is a very high-speed, very daunting Type Two turn and quite important to lap times. If you can get Turn One right, you can call yourself a driver.

Entrance to Road Atlanta's infamous Turn One. Most cars enter this late apex turn at high speed, and the bowl of the turn is somewhat obscured by elevation change. The bridge has been removed, which makes it slightly less intimidating than it once was.

Exit of Road Atlanta's Turn One. The banking and uphill exit permit relatively high speeds into the turn. A wall has recently been added at the outside of the turn to protect the trees.

will be fast if he is able to keep the car on this imaginary stripe, assuming he keeps the throttle open. Remember that it is easy to stay on the proper line for your car, and it is easy to be smooth if you are slow. Doing that while driving really fast is the tricky part.

When selecting lines through turns, use these rules: If you run out of road at the turn exit, the apex was too early. If you have road left at the turn exit, the apex was too late. Don't make the mistake of *driving* to the edge of the track if you have road left. This will give you the false impression that you are using the proper line. Just as the steering wheel must be turned at the turn in point, the wheel must also gradually be straightened from the point of throttle application, because the turn radius is beginning to increase. Also remember that the line will change slightly as you begin to go faster. Increasing your speed only four or five MPH will generate a great deal more momentum and carry your car farther to the outside of the turn.

When concentrating on your driving, remember these priorities:

1. Use the proper line around the track, not just through each individual turn.

2. Maximize traction in all three phases of the turn. Do not lock wheels in braking, don't allow oversteer to become excessive at the mid-point or later, and avoid wheelspin at the exit.

3. Regulate speed into the turn with the brakes, just enough to be able to change direction at the highest speed possible.

Learning Tracks

Regardless of what type of track you are running, you have to learn your way around it at speed. This is more than just remembering which way the turns go and which turn follows the last. In order to find and follow the proper line through a corner and around the track, a driver must know where he is in relation to reference points. A reference point is an immovable marker on the

Practice and Testing—What's the Difference?

Club racers are frequently heard at race tracks on Saturday saying something like, "Practice is in thirty minutes." And in the racing periodicals, we sometimes read of a driver testing at a particular track prior to leaving for a race at another track. Just exactly what differentiates practice from testing? An amateur racer many times needs all the time in the car he can get. It is difficult to become a good driver when the only time you are allowed to practice is for an hour or two per weekend and the race weekends are limited to a half-dozen a season. I believe that SCCA or any other group would be well advised to rent track in each Division two or three times a year just for the purpose of practice. They could charge the entrants a nominal fee, just enough to cover overhead, for use of the track for a weekend so that those competitors could practice their driving skills, try new lines, and try new setups for their cars as well. A skill cannot be perfected unless it is practiced. When the only practice a club racer gets is in qualifying and a Regional race on Saturday at an SCCA Regional/National, getting a run up the steep part of the learning curve can be difficult.

As described in Chapter 1, becoming comfortable with the skill involved in driving a racing car is more than just understanding on an intellectual level why a particular line is faster. The sense organs must be familiar with the sensations they receive during a specific activity, and the brain must be used to processing the information it receives from the senses. When both of these functions are well refined, the transfer of information from the sense organs to the brain and back again in the form of control inputs to the car is both swift and accurate. When this happens, it seems to the driver as if it is all second nature, as if he doesn't even have to think about it. When the information is transferred and processed quickly, the driver has more time to think about the abnormal situations that inevitably arise and take a disproportionate share of his time. Practice is a way to refine these functions. The more time a driver has to practice, the better his senses are at transferring this information and the more quickly his brain will become at processing it.

pavement or adjacent to it which marks your line. This can be a wiggle in the edge of the pavement, a darker patch of asphalt, a painted line, or any other permanent mark or object. Pylons and hay bales are not recommended. Although it is sometimes necessary to use reference points that are not on the track surface, it is better if these points are on the pavement so as not to direct your attention away from where you are going.

As most turns are approached, the first reference point locates the point of brake application. This marker should be exactly at the point at which you wish to begin to brake, but in the real world, it does not usually work out that way. Many times the reference point will be either a little too early or too late, and you will find yourself braking at some point *in relation to* the reference point.

By the time a driver becomes competitive on a pro level, he has logged tens of thousands of miles of practice. This cannot be accomplished in a season or two, even with an unlimited budget. It takes time for a driver to become proficient on the track. Anyone who thinks it can be done more quickly or who thinks he was born with "the gift" and does not have to work at it is deluding himself. It doesn't work that way. Even those few who have been born with the special mental and physical attributes necessary to drive a car fast and fearlessly still have a lot to learn to become successful in this highly competitive environment. Regardless of how good you are at a particular aspect of racing, someone else is always better at another. It is just not possible to be king of the hill in all respects.

Once a driver has become good enough that he can get frequent rides with pro teams, the work has just begun. Pro teams go testing to learn about a variety of engine and chassis configurations. They try different tires, shocks, they try new differentials or try an old differential in a new application, they see if the wind tunnel simulates real world conditions, study fuel mileage in differing conditions, and try new suspension geometry, just to name a few. Their cars run thousands of miles each year in conducting these tests. Sometimes a team may rent a track early in the season just to find the best setup for that track for when they race there later that year. It is a real letdown when they get to that same track later knowing that they have a great setup for it only to find it raining that weekend.

Beyond having a car that can be better set up for the conditions, an additional benefit to the driver in return for all of this activity is plenty of seat time. The driver gets a great deal of time to *practice* while the engineers are learning things! Team owners know this, and even if the engineers feel that not a lot has been learned from a testing session, the drivers are the better for it. Although there is a difference to engineers between testing and practice time, a driver should make the most of any time he gets to drive. The car may act differently when trying different things and he should note those reactions, but he should also appreciate that any type of practice means getting better at controlling the car at higher speeds. Practice may not make perfect, but it sure makes a better racing driver.

When this happens, it may be possible to use this reference point to find a sub-point, not as easily seen, which is exactly where you want to brake.

The next reference point should mark the turn in point and be followed by references marking the turn apex and the track out points. Two points make a line, but at least three are required to define a curve. These are the minimum points that mark your line. Up to a point, the more reference points you use, the easier it is to follow your line. These additional points do not have to be exactly on your line, but should show you where it is. For example, a dark spot on the pavement between the turn in point and the apex may be just to the inside of your right front wheel when you are on line. Keith Code, in his book, *A Twist of the Wrist*, uses a photographic analogy. Keith says when you have enough reference points, the scene in front of you flows like a movie. If you don't have enough reference points, the movie becomes jerky, like an old-time movie. Too few reference points will focus your attention on each too long and keep you from being smooth.

As an example of how to pick reference points, let's examine a particularly complex group of turns at the track at Hallett, Oklahoma, which is aptly named "the Bitch." The Bitch requires several reference points. As you approach the turn, at the top of third gear in many cars, on an uphill section that curves slightly to the left, the first reference point is a large, irregularly shaped dark spot about five feet from the outside edge of the pavement. A smaller dark spot accompanies it a couple of feet farther down and slightly to the right. If you sight down the lower right side of the large spot toward the smaller one and extend that line to the pavement's edge, you will notice a small white spot just on the edge of the track. That white spot is the turn in point for the first part of this three-part turn. When you turn in, you will see an asphalt seam line running parallel to your direction of travel and on that line is a small concrete patch. Aim your car so that your right front wheel passes directly over the patch. Your left front wheel should now be lined up to pass just to the outside of a concrete strip lining the inside of the turn. When you get closer to theconcrete strip you will notice "alligator bumps" on it. These concrete bumps two or three inches high are car destroyers so do not hit them. The second bump is the apex of the left hander. Put the outside of your tire at the pavement's edge adjacent to the second bump.

When you arrive at the apex, the road becomes only a horizon and does not even hint at what comes next. You will see two phone poles on the horizon just above your left front wheel or fender. Just to their right are four more poles, which are more difficult to see, which at one time supported a billboard on the Cimmarron Turnpike a half mile away. Point the center of your car directly at the center of the four poles and have faith. With your nose pointed at the poles as you crest the hill, the apex of the first right hander, which is the last alligator bump at the inside, will come up and just kiss your right front tire. Very little steering correction is required to reach the apex. At this point look for another row of alligator bumps at the outside of the turn at the exit. Aim your left front wheel at the end of that row of bumps. When you are able to distinguish the individual bumps, point your left front wheel at the third bump from the end. This is both your track out point from the first right hander and your turn in point for the second right hander.

As you turn into the second right hander you will see that the pavement is broken up and rough. This will upset the car and make it more difficult to maintain the correct line through this part of the

"The Bitch" at Hallett Motor Racing Circuit is a complex turn of a left followed by two rights. On approach, you will notice a large irregular dark spot on the pavement. This is the first reference point, which is used to find a smaller dark spot a few feet farther ahead and a couple of feet to the right.

The smaller dark spot will lead your eyes to a small white spot a few feet farther down at the outside edge of the track. This is the turn in point for the left hander. A school was being held when these photos were taken, and the white spot was marked by a small red flag. Don't count on the turn in point always being marked for you.

After turning in, guide your car to the left so that your right front wheel passes directly over the concrete patch on the asphalt seam line. Then set your sights to line up your left front wheel just to the outside of the concrete strip at the inside of the turn.

When you get closer to the concrete strip, you will notice "alligator bumps" on it. These concrete bumps two to three inches high are car destroyers, so don't hit them. The second bump, again marked by a little red flag, is the apex. Place your left front wheel adjacent to it at the pavement's edge.

When you arrive at the apex, you will see two phone poles just to your left on the horizon. At this point the track gives you no clue whatsoever as to what comes next, so you must put your faith in the poles. You will also notice four poles farther away, which supported a billboard at one time. The billboard was easier to see, but racers must adapt. Point the center of your car directly at the center of the four poles.

With the nose pointed at the poles, the apex of the first right hander will come up and just kiss the right front wheel. Little, if any, steering correction is required. When you arrive at the apex, which is the last alligator bump (marked this time by a cone), turn in and look for another row of alligator bumps at the turn exit. In many cars, you have just lifted the throttle and not yet begun to brake. Cresting the hill makes traction scarce. Be careful here.

As you exit the first right hander, aim your left front wheel at the end of the alligator strip.

When you near the end of the alligator strip, aim at the third bump from the end. It is the track out point from the first right hander as well as the turn in point for the second.

Upon arrival at the turn in point, you will notice that the pavement is very rough, which upsets the car. Turn in and aim toward the phone pole at the far right of the photo. This turn leads onto a straight, so apex it late.

Farther around the turn, you find yet another row of alligator bumps. The turn apex is the fourth bump past a large red spot on the inside of the track. The large red spot may be easy to find on Jupiter, but it may be easier to find the apex of this turn by looking straight down from the phone pole directly in front of you.

If I were driving, I would put my right front wheel directly over the small cone marking the apex. If you were behind me, it would not be there for you to use. It might be advisable, then, to remember the apex by the alligator bump closest to the racing surface, directly below the phone pole.

Centrifugal force will pull your car to the outside on the exit. Guide your left front wheel toward the end of the alligator strip.

Late apex turns make the track out point a track out area. The theoretical point is shown here by a small cone, but putting your left front at the edge of the pavement within two or three feet of it would be acceptable. These photos were taken on a cloudy day, and make it difficult to pick out reference points. The reality, though, is that races are run on cloudy days as well as sunny ones. Conditions are not always right to make it easy, but you must go fast anyway.

turn. In some cars a slight lift here will help to bring the rear of the car around and "rotate" it. This turn leads on to the back straight so it requires a late apex. Aim at the last telephone pole which is visible in the photo of this part of the turn on page 129. Farther around the turn you will find yet another row of alligator bumps lining the inside. The turn apex is the fourth bump past a large red spot on the inside of the track. The red spot is sometimes difficult to see so you may want to find the bump by looking straight down from the phone pole past the grandstands to find it. At the exit is another row of alligator bumps on the outside. You should aim your left front wheel for the end of the row. Late apex turns change track out points into track out areas because you are approaching the edge of the racing surface gradually. Putting your wheel at the edge of the pavement within two or three feet of the second bump from the end is acceptable.

Finding reference points is the key to learning a new track. One important difference between the top drivers and the rest of the field is their ability to find reference points quickly and accurately. When this ability is highly developed, a driver will be able to learn a track in only a few laps, and the reference points will flow into that "movie." Keep in mind that as you become faster, those reference points may no longer be in the correct places, and you may need to find new ones.

Keith uses a very good method of determining how well you know a track, which can be done anytime, whether at the track or at home. Use a stopwatch, sit comfortably, and close your eyes. Replay a lap of a track you think you know well from your memory. See yourself moving down the straight, entering and going through the turns using your reference points. Quite likely, the stopwatch will be off a lot from your real lap times. If this is the case, you are either not using enough reference points, or those reference points are not ingrained deeply enough in your memory, and you will be surprised by them on the track. Reacting to surprise takes time. Continue to find and remember reference points until you know the track so well that you can do this "memory lap" at the same lap times you are actually turning. You will probably never be able to do this very accurately, but it will tell you how well you know a track and assist you in learning it better.

Another method that helps is walking the track, something that was very popular a couple of decades ago. These days, time constraints seem to get in the way. At a club race, the only time this is possible is Saturday evening after the regional races (while your competitors are enjoying the beer party in the paddock). When running a pro race, it can be done on Wednesday, while the teams are parking trailers and setting up operations, or any evening after the festivities. When you walk the track, study the rough sections, pavement changes, lower traction sections of pavement, off-camber areas, places where you would not want to fall off the road, and, of course, potential reference points.

Road Courses

Road courses are like people—they come in all shapes and sizes. They are characterized by being purpose built for road racing. Simulating country roads, they have many different types of turns, and usually have something around them that it is possible to hit, such as trees, guardrails, ditches, etc. Usually located outside of metropolitan areas, they may or may not have facilities for spectators. Some of the lesser tracks are intended only for club racing. These tracks do not generally get the same kind of upkeep that the professional tracks do, and, as a result, the racing surface is generally rougher. Facilities are not as high-quality at these tracks, either. Paddock areas are often dirt or grass, and concession stands, restrooms, etc. are scarce. Professional tracks are much better in all of these areas. For a top-of-the-line track go to Mid Ohio. It is like racing on a track in the middle of a golf course.

Short road courses are generally less than 1.8 miles in length, and are usually slow tracks. Track designers seem to think that a short track deserves as many turns as the longer ones. The exception has been Lime Rock. It is 1.53 miles, and, until just recently when a chicane was added before the Climbing Turn, there was only one area where you would really slow down. Now there are two. They say it's progress.

The majority of road courses are 1.8 to 2.5 miles in length, and allow somewhat higher average speeds than the shorter tracks. Some circuits such as Road America in Elkhart Lake, Wisconsin are both long and fast. Elkhart, as it is called, is four miles long. As you should realize by now, each road course has its own character, and it is defined by elevation changes, roughness of the surface, objects surrounding the track, and so on. Get to know a track before you go to it for the first time by talking to racers who have been there before, preferably in the same type of car that you will be driving. They can help with recommended gear ratios, spring and shock rates, and alignment settings, and, as well, they might be able to give you tips on how to drive the track. Ask about specific things you should avoid and what they use as major reference points. Because of the very different nature of each road course, this type of circuit offers a bigger challenge to putting together a fast lap than do other types of circuits.

Street circuits are usually lined with unforgiving concrete, are narrow in places, and can become very congested. Learn to drive on other types of tracks. You will need all of a race driver's skills on a street circuit.

Street Circuits

Street circuits, being temporary in nature, require no upkeep when they are not being used, and more importantly, require no payments to the bank. Additionally, it is easier to get spectators to pay to see a race if they do not have to leave town to do it. Unfortunately, economics dictates that we will be running more races on street circuits and fewer on road courses.

Street circuits do not usually have a great deal of elevation change (except the old track at Long Beach), but do offer some other challenges. To drive this type of track requires great precision and buckets full of nerve. Concrete walls line both sides of the track in many areas, and always cause carnage to many cars at each race. Another hurdle is the changing track surface as you pass from one street to another. This can happen a dozen or more times per lap. Street courses are also somewhat rough. Amazingly, you can drive a street car over a race course just before it is closed off for the race and it feels perfectly smooth. Do the same thing in a race car a day later and the track will shake the steering wheel from your hands. The speed and stiff springs of a race car magnify any bumps present at all. Street circuits are by nature short and compact, and most are somewhat slow as a result. Many times the track is narrow, as well. Before you go to a street race, make sure you have developed exceptional car control. Spins cannot usually be tolerated on a street circuit. A little luck helps, too.

In contrast to street circuits, airport tracks have hundreds of yards of runoff room in most places. The concrete racing surface is hard on tires and the flat terrain and geometric turns make them less interesting to drivers.

Airport Circuits

During the Second World War, the government built training bases all over the country for air force and navy pilots. It was inevitable that racers would take them over as inexpensive road courses. Airport circuits are not as common as they were 20 years ago, but you will probably have occasion to run a few of them. Sebring is the most famous, but club races are run at many others. Airports are obviously flat. There is no elevation change at all on these tracks. Since the racing surface is 50-year-old concrete, airports are *extremely* hard on tires. They can be pretty rough, also due to their age. Where the forms were laid between sections of concrete, the tar strips cause some nasty bumps. The straights are very long, and the turns are all geometric, that is, the runways and taxiways intersect each other at 90- or 45-degree angles. You will rarely find a decreasing radius turn on an airport track. Reference points are sometimes difficult to find, so you must use some creativity. At one track years ago, the only reference point I could find for one turn was a row of 55-gallon barrels some distance to the inside. Projecting a line through them toward the track came very close to indicating the turn apex. It wasn't much, but it gave me something to aim at. The good thing about airport tracks is visibility. You can see for miles. But, basically, they are pretty boring. I would rather go to a bad road course than a good airport.

Racing in the Rain

Driving on a wet track adds two problems to the mix. One is seeing where you are going and the other is low traction. We'll deal with the vision problem first. When it is raining, the dew point is low enough that the driver's breath will fog the inside of the visor. Several helmet manufacturers sell an anti-fog liquid to put on the inside of the visor. If you wear glasses when you drive, you should also put anti-fog solution on them. Ordinary dishwashing liquid does the same thing and is much cheaper. Many of the new helmets feature ventilation systems that circulate air into the helmet to prevent fogging. A visor that is mounted on the helmet with a ratchet mechanism will also let air in when raised only slightly. Any or all of these may be used to solve the problem. A new visor, either clear or yellow (to increase contrast), should be used in the rain. An older visor that is scratched or pitted holds water and reduces clarity. Rain-X on the outside of the visor (or the windshield of a closed car) works wonders.

Rain turns everything gray, and cars ahead throw up even more spray, obscuring them in a slightly more opaque cloud. In these circumstances, it is difficult to pick out anything until you are right on top of it. Formula Cars are required to use taillights when running in the rain. Although most other cars have brake lights, none of them is required to use taillights. The logic of this rule escapes me.

Driving on a wet track requires some adjustment of line. The fast line is not the same as it is in the dry. During a race weekend, rubber is laid down through a turn by the outside tires. When it is dry, this increases traction greatly. When it rains, though, the water washes the rubber up, loosening it from the surface, and making that line treacherous. In the braking area before a turn, the wet line

Rain changes everything. The racing surface is usually the slickest on the proper dry line, requiring a change in car placement. Visibility is the other hurdle. Many remedies may be used, but all of them together do not eliminate the problem.

is about a half-car width away from the outside of the track, rather than right at the edge, as in the dry. This allows the tires to run in the areas where little rubber has been laid down. At the apex, a similar line is used, about three feet outside the normal apex, for the same reason. The rain line then rejoins the dry line at the track out point. The wet line through a particular corner may need to be modified further still to avoid puddles, oily patches, or polished sections of pavement. As always, do what is required to get the best turn exit speed.

Rain does not affect cars that normally run on street tires as much as those that use slicks. Slicks in the rain are just plain evil. They aquaplane very easily, and sometimes the driver's first warning that this is happening is the car veering off in a direction in which he is not steering. Slicks can be used almost up to the point of standing puddles on the track. After that, bring out the rains. Since rain tires are made from a much softer rubber compound than slicks, they require water to cool them. If the track begins to dry while you are out on rains, look for puddles on the straights to cool them.

In addition to changing the proper line, driving on a wet track will generally preclude the use of trail braking, too. Traction is so low in the rain that braking and turning cannot be effectively combined, and it is better to finish braking by the turn in point.

Driving on a slippery track is a very good time to hit each gear in succession when downshifting. The braking distances will be longer, giving more time to make multiple downshifts, and there is less chance of incorrect RPM matching when multiple blips are used rather than one big one. A driver should always be smooth and use gentle, flowing movements in the car, but this is especially important (and difficult) when it is wet. It is *very* easy to make mistakes. Be careful.

Ovals

The well-developed road-racing driver must be able to run ovals, too. They are not easy. It is a common misconception that all ovals are the same. We get to run two varieties: short tracks and super speedways. For Saturday-night racers, a short oval is anything up to and including a 1/2-mile track. For the rest, a short track is from 1/2-mile to one mile in length. The tracks differ from one another in turn radius, banking, and track width. Sometimes they have different radius turns at one end than the other, and even have "doglegs" in either front or rear straights. Devil's Bowl Speedway, a 3/8-mile dirt oval and favorite stop of the World of Outlaws show, has an elevated back straight which produces two climbing turns and two downhill ones. I consider ovals to be short road courses that only turn left. Super speedways are from a mile to 2 1/2 miles in length, and come in as many different configurations as the shorter ovals.

Running an oval is much like driving flat out toward a concrete wall and turning just before you run into it. The walls that line ovals are very unforgiving, just as on a street circuit, but the speeds are much higher. For example, F/2000 cars running the one-mile track at Phoenix lap at just over 126 MPH. These days, since we all watch so many Indy and NASCAR races on TV, that does not sound very fast. When you are in the car approaching a turn (and a wall) at 140 and trying not to lift though, it feels pretty fast.

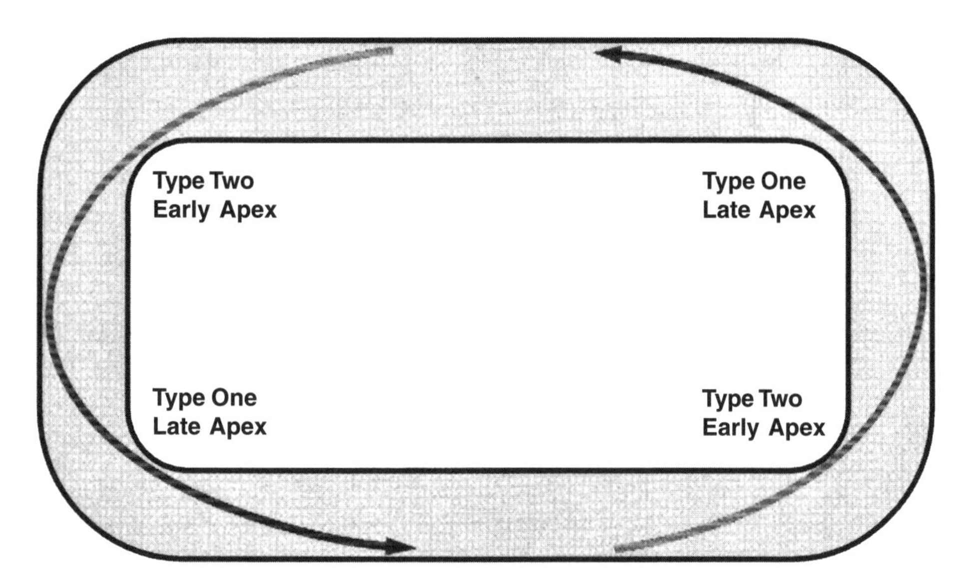

Fig. 6-7 The diamond pattern is used on a flat track or when the corners are relatively square.

The "diamond pattern" and the "round pattern" are the two lines used when running ovals. The diamond pattern is used on a track that is not banked very steeply or that has square corners at each end separated by short chutes. Indianapolis is an example. The diamond pattern turns an oval into two Type One and two Type Two turns. The round pattern is used when a track is more highly banked and when the turns are longer and rounder as at Daytona. Figures 6-7 and 6-8 illustrate these lines.

Ovals require great sensitivity of the driver. A small mistake can quickly turn into a major one, so it is very important to drive smoothly and precisely. You should avoid oversteer like the plague, but, of course, you have to drive on the limit and that usually produces oversteer. The proper setup for a car running an oval is for a slight push (understeer). This gives the driver just a bit of room for error when on the limit. When the rear of the car steps out just a bit, the oversteer must be checked *immediately* or it will turn into a spin.

If you are a bit late in the correction and it has become obvious that a spin is eminent, do not continue to steer into the spin. A spin will almost certainly end in a meeting with the wall, and it is much better to hit it with the rear of the car than the front. When you start to spin, let it go. It is commonly thought that not following this rule is what caused Gordon Smiley's death while trying to qualify for Indy in 1982. He had said earlier that week that he would break 200 MPH, like those qualified in the front, if it was the last thing he did. He was trying very hard. The roll stiffness of his car was a little low for his driving, allowing the car to roll far enough to lift the skirt of the inside tunnel, which caused oversteer. Gordon corrected for it, held the wheel to the right to catch the oversteer, and hit the wall almost head on at close to 200 MPH.

Ovals are lined with concrete, as are street circuits, but the speed and the kinetic energy are much higher. Precision driving is essential here, too.

Inexperienced oval drivers have a strong tendency to stand on the brakes hard enough to lock tires before a collision with the wall. This is precisely what should not be done. To have any control of the vehicle at all after contact, the wheels must still be rolling. Brake moderately, let the car hit the wall first, and then get on the brakes really hard, but still avoid locking tires. Any banking present on an oval will funnel cars down off the wall toward the apron. When you do hit the wall, use the steering wheel to try to keep the car up on the wall and out of the way of traffic. Turn right if you are heading in the correct direction and left if you are backward.

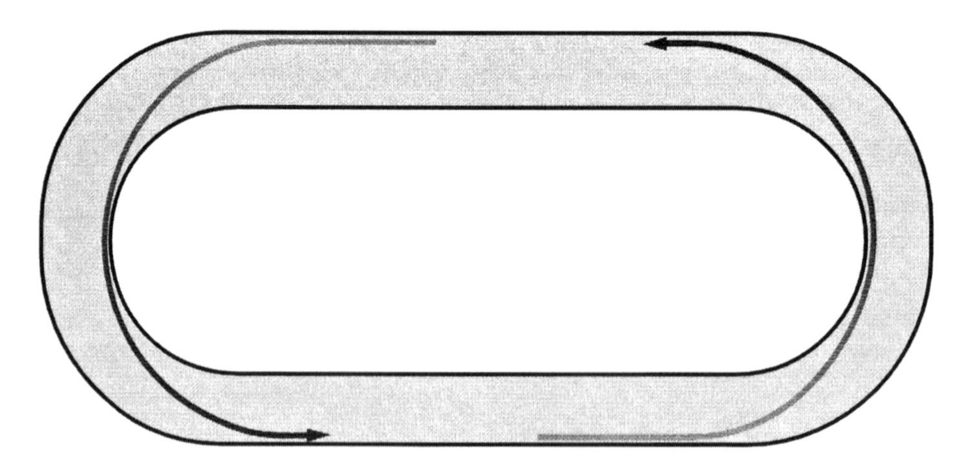

Fig. 6-8 The round pattern is used on a high banked track or where the turns are long and round. It, too, can be biased for maximum exit speed.

Develop the habit very early in your oval track career of looking around the turn to the left before or as you enter it. Any signs of trouble discovered early enough (such as tire smoke), can give you enough time to slow and adjust your line to avoid the incident. If someone spins in front of you, the rule is to aim straight for him. By the time you get there, the other car will not be. Although this method sometimes works, it involves obvious peril. The worst wreck that I ever had was in a Sprint car on a 1/2-mile paved oval. In this case, I had no alternative but to aim for the spinning car in front of me. When I arrived on the scene, he was still there, and the resulting collision totaled both cars.

Driving Big Cars

Never turn down an opportunity to drive a car. You can learn something from each of them, which you can use in other cars that you drive. When you get a chance to drive a really fast car, an Indy Car, Winston Cup, WSC, etc., an entirely new world will be opened to you. Many racers make the mistake of thinking that the big engines and high speeds make these cars difficult to drive. But the fact that they are capable of high speeds is not the reason they are difficult to drive. In reality, it is the response characteristics of the Formula and Sports Racing cars in all areas, throttle, braking and turning, and the high force levels they are capable of, which make them the province of the experienced and professional drivers. In the heavier, full-bodied cars, a great deal of momentum at speed can be very difficult to control. We have all seen the NASCAR stars spin, get airborne, tumble numerous times, and slide through the grass for hundreds of yards before coming to rest. Their weight coupled with the extreme high speeds can cause things to get out of hand in a hurry. With either of these types of car, a small mistake that would only produce a slide and a slow lap time in a smaller car, can turn into a major problem. When you get an opportunity to drive one of these cars remember that they may be a bit different than the smaller cars that you are accustomed to. Take it slow and easy until you get the feel of it. High speeds greatly magnify the car's weight, and in a responsive car, a small control input mistake can be magnified as well.

Mental Driving

Timing

Timing is everything—Ziggy

I once saw a cartoon in which the character, Ziggy, was watering his lawn with a hose—during a rainstorm! The caption said, "Timing is everything." Driving a race car well is all about timing. If your timing is not perfect, you will not be fast. As are most things involved in racing, timing is more complex than just standing on the brake pedal at the right time. At 180 MPH you are traveling 26 feet every 1/10 second! Average reaction times are in the .07 to .09 second range, which means that the average driver has traveled 18 to 23 feet, more than a car length, while the message was traveling from his brain to his limbs. Fast reflexes and good hand-foot-eye coordination are involved in timing, but they are no substitute for being able to decide exactly when to initiate an action. Timing is understanding what should happen when and where; it is not just reacting quickly.

Timing is temporal as well as spatial. You are not just dealing with linear time, but a relationship between time and position on the track. Your timing, therefore, is not determined by when to act as much as where. A single action can be easily timed or, rather, *spaced* in relation to the track, but since many actions are required just to get through one turn, the actions are related to each other by time. It has become customary to group all of these actions and talk about their timing. You should remember, though, that the important relationships are between the actions and *where* they happen.

Timing involves anticipation. Years ago, when I was running Karts, a friend invited me to go with him to the first test of his new Sprint Kart. When my time to drive came, I was amazed to be running off of the *inside* of some of the turns! The track was familiar, and I was turning in at the points I knew to work in my Kart. Upon reflection, it became clear that my Kart had a corner entry understeer problem and the new Kart did not. Without realizing it, I had been anticipating the distance required for my Kart to begin to turn after I turned the wheel and applying the same anticipation to a Kart that did not need it. My timing was incorrect.

Here is another slightly more complex example of timing. Suppose you are going slightly uphill, nearing the end of the straight. The hill crests and the road falls out from under you just about the time you would like to get on the brakes for the turn. When the tires are unloaded, cresting the hill, they have very little traction, and it would be very easy to lock a tire. But if you wait until you again have good traction to apply the brakes, your braking might be too late. Your braking point must coincide with the point at which the car comes back down and regains load. With practice, your timing can become good enough that you actually begin to apply the brakes while the tires are unloaded, and are on the pedal hard precisely when the car comes down and has the suspension loaded the most, generating the highest braking force. In this way, the brakes can be used to best advantage to slow the car just enough to use the proper line through the turn.

Of course, this is only one timing event. In the real world, this brake application might be followed by a hairpin turn that is slightly banked going in, has a decreasing radius, has a bump just before the apex and falls downhill, and is off-camber coming out. This section of track requires many timing points. When you get each one of them just right, it feels as though you have done everything in the proper *rhythm.*

Concentration

> *To have something other than driving on your mind while driving nearly 200 MPH is called suicide. Total concentration on what you are doing is not only a recommendation for successful racing, it is a necessity for survival.*—Mark Martin

Meditation. Yes, I know what you are thinking: "What does meditation have to do with driving a race car?" The reality is that if you are driving a race car properly it *is* meditation. An observation with which many agree is that *most people drive like idiots*. If you live in a large- or medium-sized city, you are probably rarely able to drive very far without encountering one or more drivers who make left turns in front of you, change into your lane without looking, or commit one of a dozen other offenses that endanger your life or property. These actions, and the accidents they often cause, are almost always a result of one or both drivers not paying attention. When one of these inattentive street drivers does something stupid in front of you, it is likely that he is thinking about the deplorable conditions at the office, or the problem with Aunt Martha, or something else that takes his attention away from the moment. To drive a car well, whether on the track or street, takes awareness and experience of the moment. These are sometimes combined and called simply "concentration." Driving a race car takes so much more concentration than anything else you have done, that it is imperative to learn how to master pure concentration or *samadhi,* translated from Sanskrit as "concentration, the quality of meditation."

As used in the Eastern world, meditation is an act of concentrating on nothing but the present moment. Most practitioners of this discipline advocate some form of a relaxed posture in a quiet room while focusing on nothing at all. As thoughts about Aunt Martha enter the mind, they are acknowledged, but let go without attachment. With practice, one is able to meditate with enough experience of the present moment that extraneous thoughts do not intrude.

It is not necessary to sit in a darkened room and recite mantras to drive fast. In fact, meditation can be and is practiced in many different ways. Have you ever stared at a campfire or fireplace in silence? Amazingly, most thoughts disappear of their own accord at these times, and you are left with only awareness of the fire and experience of the present moment. Dan Millman, author of several self-improvement books and conductor of seminars on the subject says this, "Meditation practice doesn't always involve sitting down at all. We can meditate standing up, lying flat, or moving around—as long as we stay mindful and maintain the witness position, allowing our attention to focus on something other than our thoughts. As we walk, or engage in a sport or other form of movement, we can practice dynamic, moving meditation. We can also do this by paying attention to what we are doing as we drive a car—*especially on a race course where our attention is less likely to wander* [emphasis added]."

When I was a novice driver, while on the track I constantly thought about all the things I had been taught to do. Turn in at the right point, feel the balance of the car, rev high enough when down shifting, etc. Thoughts ran through my mind in a never-ending stream. Sometimes I would actually shout to myself inside my helmet something like, "Boy, you really screwed up *that* turn!" It wasn't until much later, when I was able to calm the noise inside my mind, that I was able to get down to the job of driving the car fast.

All thoughts are rooted in either past or future. When you have a thought about something, it is about how it looked, felt, etc., in the past or what you would like to do about it in the future. Even the thought you have right now is about the words you have just read. It is impossible to think about the present moment. As soon as you do, it becomes the past. In a race car, a driver must continuously process a steady stream of information, and if he tries to think about what he is doing, he is either thinking of what he just experienced or he is making plans for how he will deal with the coming situation.

To be fast, total concentration is essential. The ability to eliminate all extraneous thoughts and focus only on the road in front of you is something that must be learned. Brian Goellnicht, driving a March-Chevy in this photo, is a master.

Some planning is obviously necessary. For example, as you approach a turn, you must select a time to brake, downshift and turn in. If you have picked the proper reference points, though, these things will happen almost on their own, requiring very little thought about them. It is not necessary at all to think about what you just experienced. As a matter of fact, such thoughts are distracting. Learn to focus only on the scene in front of you and the sensations that tell you what the car is doing at the moment. Use all of your senses: take in the vibration in the steering wheel, the pitch of the engine, and the G forces acting on your body. But do not make value judgements. Learn from your mistakes without dwelling on them. In the movie, *Karate Kid*, while teaching his young student to drive a nail into a board with a single hammer blow, the old Zen Master says, "Think only nail. Nothing else, only nail."

In the beginning, I made a lot of mistakes. Thinking they were the result of unpolished skills, I deemed them normal for my stage of driver development. In hindsight, though, I realize that the multitude of thoughts of what I was supposed to do, and value judgements I made after trying to do them, were taking my attention away from the task at hand, which was driving the car in the present moment. Sure, there are many things a beginning driver must learn to do, but at some point in his career, these things must become second nature and the focus must become road. Nothing else, only road. The sooner you reach that point, the faster you will progress.

Making Decisions

> *In racing, you are creating an almost continuous emergency situation by pushing to your limits.*—Keith Code

You will probably never be able to reach the plateau of driving totally in the present moment, and the reason is simple: You will always be learning. Because you are always trying to improve your own performance and to outreach your competitors' improving performances, you must continually try new strategies and techniques, and this takes thought and attention and decisions. Everything you do in a race car requires either a decision that has been made as a result of your prior experience or a new decision to be made in the present.

One method of making decisions in driving a track is to work away at a problem, and through a process of elimination, arrive at an action that works best. This method is slow and laborious and provides much possibility of making mistakes, since you will be trying some things that may not work at all. In time, though, you will have tried so many different techniques that you will have built up a library of actions, each of which has different benefits and can be used in particular situations. This is the benefit of experience.

Seat-of-the-pants drivers usually rely on the trial and error method. To improve quicker, though, a driver should become a thinking driver. To do this, when you are confronted with a particular problem on the track, you should think it through to determine what is causing it, and then determine what might be done to correct it. Price Cobb ran Super Vee in 1983 with Carroll Smith as his engineer. During the first half of the season, when they had a testing budget, Carroll would make adjustments to the car to verify an assumption and Price would modify his driving accordingly to make up for the problem created by Carroll's adjustment. By doing this, however, Price was rendering the test session useless, as Carroll was not able to find out what he wanted to know. They got it straight after a bit of initial wasted time. The point is that Price thought through the handling problem he had on the track and corrected it by reasoning. He is very much a thinking driver.

Michael Roe won the Can-Am championship in 1984, and is a blindlingly fast driver. Michael relies on experience to tell him what action might be best in a given situation. He is just as fast as Price when everything is right. When he has a problem, though, he has to go through his mental library and try a couple of actions to find a solution to the problem. This may take couple of laps. Meanwhile, Price might have solved the problem analytically and pulled out a lead.

The best approach seems to be a combination of the trial and error method and some analytical thought. Don't be afraid to try different techniques on the track, but supplement them with study of vehicle dynamics, and think about how you can make the car do what you want it to do. A basic example is a car that understeers too much. On pit road, many adjustments can be made to remedy this problem, but on the track, you must do it by driving the car. To compensate for understeering, you can dive in a little deeper and stay on the brakes a little longer while turning the car to bring the back end out. This is known as pitching the car into the turn. Speaking of his

Michael Roe was brilliant in the VDS Can-Am. The fact that he drove by the seat of his pants proves that no one way is correct.

early racing days, Foyt explains, "You had to develop a skill with cars that didn't have the first bolt of sophistication. You *made* them into race cars, yourself, right there on the track." In those days, cars did not have the sophisticated adjustments that they do today to dial them in to the prevailing conditions. In order to make an ill-handling car handle properly, A.J. had to change his driving style. Foyt has a lot of experience, but he thinks about what he is doing with a race car, too.

Many times, due to chassis improvements or your progress on the learning curve, you will find that a maneuver or a method that has become your norm no longer works. When this happens, it is important for you to completely erase the original method from your library and replace it with a new one. In an emergency situation, you will go back to the method you know best, the one most deeply ingrained in your mental library. Unless it is erased completely and replaced with the new one, it is likely that you will pick the wrong method if you have little time for decision. An example might be a race at a track you have run many times before, but in a different kind of car. The new car will probably feel very different and not react in the same way, causing you to change your driving style. Unless you are totally consumed with driving the new car, you may try to drive it as you would your old one in an emergency situation. It's a little like calling your wife by your ex's name. We all know how dangerous *that* can be.

Sensitivity

Feel the track with the tires as if they are your hands and feet.

Most street cars these days are very comfortable. They use thickly padded leather upholstered seats, power brakes, power steering, fluid-coupled automatic transmissions, and suspension and steering designed to isolate the driver from the road. This is all well and good if you are taking a road trip or are going out for a romantic evening, but these amenities interrupt the flow of information from the car to the driver. Showroom Stock notwithstanding, we don't race street cars. Race cars are designed and built to maximize the flow of information from the car to the driver so the driver is more easily and quickly able to make the necessary corrections to keep the car on the edge of the traction circle. This, in part, is the reason for the car's hard, non-padded seat, suspension mounted to the chassis without rubber isolation bushings, very firm brake pedal, and the like.

Once while driving a Kart, I was going through a twisty bit of track that was very rough. Not having suspension, Karts would feel pretty rough on a billiard table. It was bouncing around, and I was trying to run the proper line, and in the middle of everything, I felt a bump from the left rear and immediately saw a small black object fly over my left shoulder. I thought the left rear tire had chunked. I pulled into the pits and found nothing wrong—except the little rubber pad had come off of the brake pedal. It had fallen onto the track, been run over by the left rear, and launched over my shoulder. I was wrong in my initial interpretation of the event, but I could hardly believe that I had actually felt it when I ran over such a small, soft object among all of the other bumps in that area and picked out that one bump as significant.

It seems amazing that it is even possible for a driver to be sensitive to the subtle information he receives from the car in such a hostile environment. The force levels are so high amid the banging and crashing, cornering and braking, noise, vibration, and the rest, that it would seem that no one would be able to pick out something as minor as that bump from a rear wheel. But we can. A human being is remarkably efficient at gathering information from his surroundings, and that is a very good thing. Sensitivity is a very important trait for a racing driver. It sounds corny, but you should learn to feel the track with your tires as if they were your hands and feet. Practice it by being aware of all the subtleties of driving your street car.

Smoothness

Drive smoothly, but don't use smoothness as an excuse to drive slowly.

No discussion of the fine points of driving would be complete without mentioning the importance of smoothness. As noted earlier, tires do not respond well to abrupt changes in the load placed on them. In order to keep from upsetting them and losing traction, it is imperative that the car be driven smoothly. This means using slow, controlled, planned, fluid movements with the controls. You must turn in gently, not jerk the wheel. Shifting should be done smoothly including RPM matching. Braking should not be done by standing the car on its nose. In addition to maximizing traction, driving smoothly will let the driver use his senses to notice subtleties of the car's behavior. Since the car is to be driven right on the limit of traction, it is necessary to drive smoothly to be able to discern the small nuances of attitude. Throwing the car around masks these.

On the other hand, when the car is not properly set up, you should be prepared to use any tactics necessary to make it do what you want it to. If it understeers going into a turn, for example, you may need to "pitch" it to get it pointed in. In his autobiography, A.J. Foyt says, "But I did learn that most basic lesson: If a car doesn't handle you *make* it handle." He is not referring to making chassis adjustments; he means doing the things on the track that are necessary to get the car to work. Be smooth as much as possible, but do not use smoothness as an excuse not to make the car do what it should.

To Anticipate or Not To Anticipate

> *Using anticipation incorrectly can lead to too many corrections and will impede smoothness.*

Drivers sometimes say that to drive fast, you must anticipate what the car will do and correct for it before it happens. They are usually inexperienced; this method is highly overrated. The time to use anticipation is when you know the car's reaction to a specific occurrence such as a certain bump. By knowing how the car will react when it hits the bump, you can correct for it just *as it happens.* Unless you are expecting a specific reaction to a particular condition, though, you should wait for those little nuances of the car's behavior, such as a little twitch of the rear, and correct for it with reactions quick enough to catch it as it happens. This is when quick reactions are essential. While there are times to use anticipation, a driver who uses it indiscriminately will inevitably be jerky and will make two or more corrections when only one was required.

If you have raced Karts, you know that they are *extremely* responsive. A Kart responds immediately to either control inputs or external forces. This demands that the driver's reactions be very good indeed and is the primary reason why Karts are such good training for bigger cars. In a larger car, due to a higher center of gravity, higher polar moment of inertia, longer wheelbase, and a host of other design parameters, the response is much slower. The driver's reactions have to be just as quick, though. Although things happen much slower in a car, it also takes more time to gather up an out-of-shape car. The action to catch the rear of a car coming around must be initiated at the very instant it is felt. If you react too late, you will probably have to make larger corrections to catch it. At best, this will cause more scrubbing of the tires and slow the car. In the worst case, you will be a step behind all the way, and the result may be a spin. Be smooth and gentle and plan your actions, but also ad lib a bit to correct for the little things.

Breathing

The car missed the wall and spun the other way, down into the infield. It was then that I realized I had been holding my breath. I guess I always held my breath when I was sliding, but I didn't realize it until then.—A.J. Foyt

When I get into a tight situation on the track, I hold my breath. This also happens while welding in a hard-to-get-to spot. When you hold your breath, it is because you are tense. When you are tense, your muscles become tight and your movements jerky. Precise car control is very difficult if you are tense, and that tends to happen when you most need precise car control. When faced with an emergency situation, all of your attention will be used in trying to get out of the situation. You will have very little left over to think about relaxing. These situations don't happen with lots of warning or they wouldn't be emergencies. It is difficult to prepare for what you will do when the time comes and the adrenaline begins to flow. I have not yet mastered relaxation in these circumstances, but I am doing better. I practice while I'm welding as well as driving. Someday I hope to get out of a close call and realize that I *did not* hold my breath!

Road Blocks

Road blocks are a driver's version of writer's block. Don't be frustrated by them, but view them as doors to the next level of lap times.

Each time you return to a track, the goal should be to go faster than you did the time before. Especially in the beginning stages, this is relatively easy. Times will occur, though, when you seem to come up against a lap time barrier that is tough to break through. When this happens, it is a signal that you have a problem area that you need to work on. There are two methods of discovering which part of the track is causing the problem.

The first method requires that you take note as you are driving on the area of the track that seems to be compressed the most. The apparent compression occurs as your driving proficiency increases because, as your speed increases, you have less time available to perform the required car control actions. The reference points become timed closer together and the whole track seems to become compressed. If you are not mentally prepared for this, you will be driving a bit over your head. The section where you seem to be rushed is the one you should work on first in order to improve lap times.

The second method requires that you again run a "memory lap" and imagine yourself driving around the track. If there is an area that seems a little fuzzy in memory, it is a problem area as well. You may find that the areas you cannot remember as well coincide with the ones where you seem to run out of time.

After you have determined *where* your problem areas are, you need to find out *what* the problems are. Each situation is different, but here are some possibilities. The most likely is a lack of accurate reference points. If your reference points do not seem to be in exactly the right places or if you cannot find enough of them, it's time to walk the track. Even driving around at 30 MPH in a street car can be beneficial. Another potential problem may be that you are going into a turn too hot and trying to crowd too many activities into too short a span of time, creating a situation in which too much of your attention is used trying to save you own life. If this is the case, take some time to analyze the situation and decide where you can slow down in order to increase control and precision. Or perhaps your timing of those activities is not precise enough. Even though you have accurate reference points, you may not be timing your actions to them accurately. Another cause may be that your line takes you through a rough section or across a bump which upsets the car enough to require additional actions. After you determine the reason for the problem you are having, correcting it may be relatively easy.

Just because you overcome a particular road block, does not mean that you will not have another one or that you will not come up against the same one again. As you go faster, you will continually encounter lap time plateaus. Don't be frustrated, but view them as doors to the next higher levels.

Chapter 7

Learning to Race

Always use the piece of road the other guy wants.

Qualifying

The starting order for a race has been determined by a number of methods over the years, but using lap times from qualifying sessions has become almost universal. It is important for a driver to learn to qualify well. Starting position is more important on some tracks and in some races than others, but it is never unimportant. In a race like the 24 Hours of Le Mans, more time may be available to pass competitors, but after the field has spread, it will still be more difficult than in qualifying. On a track where passing is easy, it is still important to qualify well, but when a track makes passing difficult or even if only the first few turns are difficult places to pass, it is especially important to be as high on the grid as possible. The more of your competition you can pass before the race starts, the fewer cars you will have to pass after it starts when it is more difficult. Many pro races pay a bonus to the pole sitter, too.

Qualifying is done in two ways depending on the sanctioning body. Single car qualifying is usually used on oval tracks. With this method, the driver must be able to turn on a burst of speed on command. You will only have two to five laps on the track. Use your first lap to get the tires, brakes, and drive train up to temperature. A race car is not built to run cold. When you take the green flag, everything should be hot including your determination. This is the time to drive flat out at ten tenths. Few drivers have ever been able to do this for more than a few laps at a time without making a catastrophic mistake. Most drivers must build up to this kind of driving slowly, over a number of laps, and are not good at single car qualifying. Even when using lengthy multiple car qualifying sessions, many drivers are faster in the race than in qualifying. If you can develop the ability to turn on qualifying speed like a switch (without making mistakes), you will be able to start ahead of most of your competition.

Multiple car qualifying is more dangerous than qualifying alone because of the need to go really quickly in traffic. However, schedules do not usually permit single car qualifying at club races and many pro road races because of the number of classes running. When qualifying in traffic, you must be able to judge when to "turn it on" to avoid the traffic which can slow you down. Many drivers try to go fast for the entire session, only to get caught up in racing with other cars. In an effort to get by to get a clean lap, they waste the entire session. It is better when running up on slower cars to drop back and get some running room so that one complete fast lap can be turned in. Only your fastest lap will be used to determine your starting position, but it only takes one car getting in the way the way to waste it. And it only takes one mistake on your part to waste it, too. In multiple car sessions, you never know when a yellow flag will come out, or when someone will blow an engine and oil down the track. It is best to get your fast lap in early in the session and also in the first session if you get to run more than one.

Racing demands that you prioritize your attention to different areas, and the priorities will change depending on prevailing circumstances. In qualifying, three areas need to take top billing: controlling your line, maximizing traction, and regulating speed. In controlling your line, use at least the minimum reference points as discussed in the last chapter, and use the line that will minimize lap times. Saving tires and reducing load on the car are not priorities at this point.

Speed will be highest through and out of a turn if you maximize traction. To do this, you will need to program your "biological computer" to use the sensory data you receive through your hands, feet, seat of pants, and middle ear concerning G loadings, both longitudinal and lateral. This will allow you to notice in minute detail how much of the available traction the car is using at a given instant. This is an area where track time yields large returns. By driving lap after lap, your brain is slowly programmed to sense the subtle variations in attitude of the car and traction of individual tires. Speed and line directly affect the traction that the tires afford at any given point in a turn.

Braking is only one part of regulating speed. The throttle must also be used, smoothly, to attain the maximum speed possible through and out of a turn without generating wheelspin and/or oversteer.

Starts

It should go without saying that you will want to get the best start possible. Any cars that you can get by before the first turn are cars that you do not have to pass when it is more difficult. You will only encounter two kinds of starting procedures, standing starts and rolling starts. Of these, standing starts are more exciting, more challenging, more equitable—and more dangerous. Only Shifter Karts use standing starts in this country, now that they have been dropped from pro Formula Atlantic, but standing starts are the norm in many international formulae such as Formula One. A standing start is a drag race to the first corner. Knowing how long the flagman takes to drop the flag or how long it takes the green light to come on is crucial to a good start. Find out how it is done at each track. The flag or light must be anticipated along with your own reaction time. If you are planning to run a class that uses standing starts, spend some time studying the way the professional drag racers cut the lights.

When a start is imminent, bring the RPM up to just under red line and hold it. Do not rev the engine up and down while waiting for the start. This is a good way to find the RPM dropping when you get the green. Watch the light or flag. When the time comes, let out the clutch pedal quickly but gradually, and slip the clutch slightly while bringing the slip rate of the tires up to the point at which maximum acceleration is generated. You will know when this point is reached because the rear tires will be barely slipping, and the rear of the car will be slithering sideways just slightly requiring steering corrections to keep it straight. Don't get carried away with this or you may find yourself slow, or worse yet, sideways. Be careful not to slip the clutch too much or you will be retiring early. The correct balance point of clutch slippage and tire slippage must be found. By the time you are one-half to two-thirds through first gear, the clutch should be completely engaged. This point will be reached very quickly as will the one-to-two shift. Practice starts in your private testing sessions, but only when you have a need to learn standing starts. In many cars the drivetrain is not built to handle these severe loads.

Unless you are on the front row, the danger is in getting together with a car in front of or beside you. Expect it. Be ready for the car beside you to move over on you, and expect that the car in front will not accelerate as fast as you (or at all). Standing starts require a great deal of concentration and awareness. Your acceleration can be controlled by the car in front. If this happens, try to pull out to the side to get some running room. Remember first that you must have some room beside you, and second, that other drivers will be trying this technique, too. Standing starts can get pretty crowded, and things happen quickly.

Rolling starts are less dangerous, but are not without their own difficulties. The starter likes to see a well-formed grouping as the field approaches the flag. If he does not see this, he will send the field around again for a second try. The reason that the field would not be well formed, of course, is that everyone wants to jump the start. Multiple pace laps are frustrating and time-consuming, and they use fuel. If you are not on the front row, stay in line as you approach the starter. Drop back just a bit so that your speed is not controlled by the car in front of you. The field will increase speed anticipating a start. Watch the starter, not the other cars. When the flag drops, you should have "anticipated" the event and be well into passing the car in front of you with a speed advantage. Watching the way the flagman starts other races can give you a clue as to how he will start yours. The speed of the field, as it comes around to the flag, is controlled by the fastest qualifier. He can do this at any pace he chooses, and races have been started at anything from a snail's pace to almost racing speeds.

If you have qualified on the pole, you have the responsibility (and advantage) of bringing the field around at your pace. Price Cobb started a race at Elkhart Lake once and brought them up the hill toward the flagman at a moderate pace. As they approached the flag, he slowed a bit, requiring the other cars to slow also, and as the domino effect caught up with everyone, Price sped up. About that time the flag came out. He got a hell of a start and went into the first turn about five lengths ahead.

Regardless of the type of start, the first turn is the most dangerous of the race. With everyone bunched up and adrenaline flowing, accidents are common. At this time, you should not give up any ground, but you should also not do anything that may cause an accident. Give other cars running room, but just barely. Remember the basic racing rule: you want to be using the piece of road the other guy wants. More races have been lost in the first turn than have ever been won there.

Depending on the quality and size of the field and how tight the track is, the first two or three laps can be almost as dicey as the first turn. During these laps, everyone will get sorted out, and the turns will progress from being groups of cars all going for a corner at once to short strings going through in single file. Until this sorting procedure is complete, most of your attention will be focused on racing, not on finding the proper line, or even on checking instruments.

If you find yourself in the lead on the first lap, you know the track is clear and you can put in one lap as if it were a qualifier to try to stretch out your lead. When you return to the start/finish line, however, you should slow to racing speed. There is the possibility that a first lap incident has left cars on or at the side of the track, or has left debris on it. It is also possible that someone could have dropped a liquid on the track or thrown dirt on it. You should be prepared for these things during the race, but on the first lap, you know it is clear.

Offensive Passing Maneuvers

Drafting

Drafting, or slipstreaming, as it is sometimes called, is the easiest and most effective passing technique, if it is done properly. When a car is traveling at high speed down a straight, it cuts a "hole in the air" behind it, in which the local air pressure is reduced. The size and magnitude of this low-pressure area is a measure of how much aerodynamic drag the car is producing. If a following car pulls into this area, his drag is reduced because he does not have to push his car through as much air as his competitor in front does. At the same time, this will reduce the drag of the front car by deflecting some of the air passing over and beside it into a longer, more aerodynamic shape. Two or more cars drafting each other down a straight can actually go faster than any one of them alone. For this reason, cooperation among drivers can allow them to pull away from a group.

The second car in line has an advantage in that he can use a smaller amount of throttle to maintain his speed because of the reduced aerodynamic drag. He can use the extra throttle available toward the end of the straight to pull out of line and pass the leading car. Doing so will increase his drag, but he will be able to use more horsepower to overcome it. (The drag of the leading car will increase, too, but this car will already be operating at full throttle.) This technique must be used close enough to the turn that the overtaken car does not have sufficient room to re-pass, and must also be used far enough away from the turn that there is time to make a clean pass. If this is done correctly, the overtaken driver can do nothing to avoid being passed. Figure 7-1 shows how the drag reduction works.

When drafting, engine temperatures will rise because of the reduced airflow through the radiators and coolers. This will only be a problem if the drafting continues for several laps. If this is the case, it will be a good idea to either drop back just a bit or to pull out to the side to allow the engine to cool before continuing the attack. Aerodynamic downforce will also be reduced when drafting, and this is important if you are driving through long sweeping high-speed bends, either on ovals or road courses. This is a dangerous situation if your car normally creates substantial downforce.

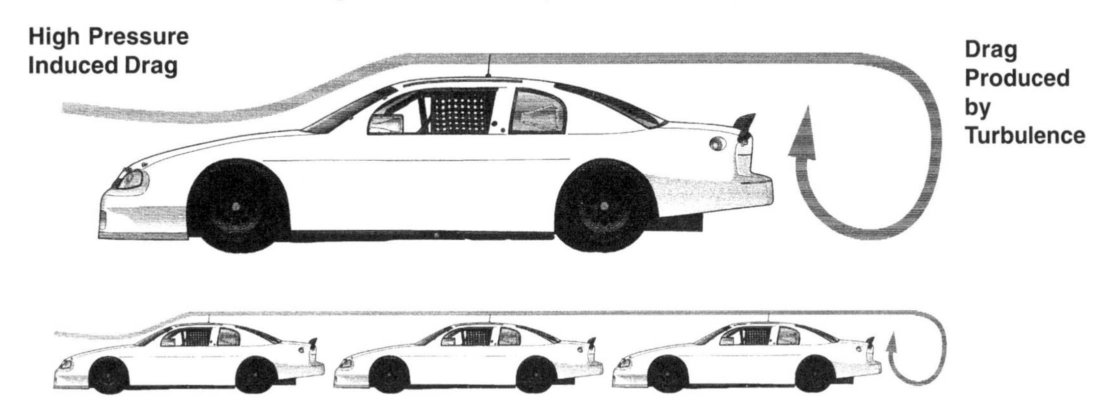

Fig. 7-1 In a multiple car drafting situation, the drag of the front car is reduced by the second car filling up the "hole in the air" behind it. The second car has neither the high pressure area ahead of it nor the low pressure area behind it. The third car does not have the high pressure area ahead of it, but does have the low pressure area behind it.

Other Offensive Tactics

When you are qualifying or running alone, you should be trying to beat the track better than the other competitors, but in a racing situation, this does not hold true. If someone is in front of you, your job is to get around him any way you can without touching him. This is contrary to the philosophy of most oval track racers, which seems to be, "Knock 'em out of the way if you have to." I once drove a Sprint car for an owner who was constantly telling me that I shouldn't have a straight front bumper. He would say, "Use that thing; that's what it is there for." Road racing, especially in open wheeled cars, teaches you just how close to someone you can run without getting into trouble. If you have to "knock 'em out of the way," it is not a fair fight.

This should not be confused with unintentionally hitting someone. At a recent Indy Car race in Portland, Michael Andretti inadvertently hit Brian Herta from behind. Herta was put out of the race, while Andretti continued. Herta was understandably upset and said that Michael had misjudged his own speed into the corner. Upon seeing the incident from Andretti's perspective, however, it appeared that Andretti had misjudged Herta's speed, which caused the incident.

When lapping someone, your speed is likely to be sufficiently higher that passing is not a problem, so long as he knows which side you will pass on and approximately when. He should be using his mirrors and wave you by. Don't count on it. Drivers who will be lapped are often slow because they are in a little over their heads. Consequently, they are paying more attention to where they are going and other driving functions, and less to watching their mirrors as they should. When lapping a driver who has not waved you by, give him a wide berth. If you will be passing him in the braking zone, pull in behind him briefly before pulling back out to pass so that he will know you are there and not turn in on you. A slight change in someone's mirrors will get his attention if it happens close enough to him. If he has waved you by, though, you are not required to pass on the side he indicates or at the time he indicates. If you pass at another time, give him some extra room then, too.

Passing someone who is just a bit slower than you are can sometimes be tough. Almost always, the best way to do this is to worry him into making a mistake, which will give you room to get by. "Filling up his mirrors" makes him drive defensively. He, too, will be trying to use the piece of road that you want. By convincing him that you want a patch of it other than where you really do, you can move him away and give yourself room to pass. Many drivers like to use this technique to act as if they are trying to pass on the outside in a given turn for a couple of laps. After these feigns, the leading driver will move up to protect his line, and the overtaking driver dives down underneath him. Other times, just being on another driver's tail for a couple of laps is enough to cause him to make a mistake. When that happens, you must be ready. Foyt explains it this way. "You've got to think fast. The second you see a hole or the instant the guy in front of you has a slight case of brain fade—say, he just hesitates a second—well, you have to make your move immediately, and it has to be the right one, because you don't get a second chance. Either the hole fills up, or if you've made the wrong decision, you screw up."

Sometimes, you must really work to make a pass. If the other driver is just about the same speed, it may take a couple of laps just to work yourself into the right position to even challenge him. If you make a very slight mistake and drop back a couple of car lengths, it may be two or three more laps before you can make it up again and resume the fight. These battles are the ones that you will both remember. While you are behind him, study his lines to see where he is good and where he has trouble. Pull out all the stops and use your best psychology on him. To make this pass, you must go at it like you are both a couple of U-boat captains struggling for survival. Remember, though, that while all of this is going on, you will be slowing each other down. This may allow other drivers to gain on you.

During a recent Formula car battle between two experienced drivers, one which lasted several laps, a passing maneuver was seen which amazed everyone on pit road. The overtaking driver was using a very wide line onto the straight to maximize his speed for a pass in the braking area, a strategy that had failed for several laps. Finally he drifted out onto the straight and then pulled to the inside of the track. The leading driver, sensing that something was awry and wanting to protect his position, pulled to the inside also and continued down the straight. At the last moment, the overtaking driver swung back to the outside and took his normal line into the turn at the end of the straight. It was too late for the leading driver to do anything about it, and he had to brake early since he was using a very tight line. This pass was completed by Al Unser, Jr. over Jacques Villeneuve, and was one of the most brilliant passing tactics anyone had seen for a long time.

If your competitor has a problem at the exit of a turn that is followed by a straight, the pass will be relatively easy. Drop back just a bit so that the position of his car is not controlling your speed through the turn and take your best line to maximize speed out of it. The pass will be completed on the straight, or, at least, it will give you the speed to pull alongside in the braking area for the next turn. If your competitor has trouble at the entrance to a turn either in braking or due to an understeering car, just try to pull inside in the braking area , which will cause him to slow or take a wider line. Either of these will give you a chance to pass.

Many novice drivers have tried banzai passes by pulling alongside another car and early apexing a turn that should be late apexed. This will give the passing car the lead momentarily, but it is so slow at the exit that the other car can easily resume the lead. This method never works. Don't fool yourself into thinking it will.

Here are some fundamental rules for passing:

1. **Drive your own race.** Don't glue your attention to your opponent's bumper or gearbox. In this way, you can pick your own lines rather than remaking your opponent's mistakes.

2. **Never let your competitor know where you are faster than he is.** Keeping him in the dark gives you the element of surprise.

3. **Think fast and be decisive.** When you make a move, first make sure it is the right one, and then be prepared to follow it through.

4. **When passing on the inside, the corner is not yours until the nose of your car is alongside your competitor's front wheels.** If you are not that far forward, expect to have the door closed.

5. **When passing on the outside, the turn is not yours until the nose of your car is ahead of your competitor's.** You can run low, however, and force him to slow, allowing you to get by in the braking area or at the exit. If you do this often, get a fiberglass sponsor.

Defensive Passing Maneuvers

Defending your position from a competitor is more difficult than planning a pass because you cannot make the moves; instead, you must react to his assaults. In order to do this effectively, you should always know where he is. You cannot use the piece of road he wants if you don't know where that piece of road is. You can keep up with his movements by watching your mirrors and by listening to his engine noise. The act of taking up space where your competitor wants to be is called "making your car wide," but it is also sometimes known as blocking.

This is a somewhat dubious term. Some may remember Al Unser, Jr.'s first race at Indy where he "blocked" Fittipaldi in an effort to help his father win. Junior was not racing with Emmo for position, but his father was, and Jr.'s tactic let Al Senior pull away. This did seem like intentional blocking to many, and, as such, is not recommended. If he had been racing with Fittipaldi for position, however, it might have been perfectly acceptable. Bear in mind, though, that since everyone seems to have a slightly different definition of what is acceptable as defensive racing tactics, and what constitutes intentional blocking, protests from competitors do occasionally occur. The consensus seems to be that a competitor can make one move off the normal line to defend his position, but if it is closely followed by another, it is blocking.

When a competitor is trying to get around, he will be looking for your weakness to capitalize on it. Since he is in a position to watch you more closely, he is likely to know more about how your car works than you do about his. Pay attention to where he is and where he is trying hardest to pass, but do not pay so much attention to his maneuvers that you neglect your own driving. Doing so is almost certain to produce a mistake on your part which will let him by. Disregarding one's own driving is the most common mistake a driver makes when being hotly pursued.

Some drivers have been known to drop a couple of wheels off the inside of a turn to throw some dirt and rocks at the following driver. Many times, this will make him back off for an instant. Because of our conditioning to slow when we see red lights ahead of us, it may be possible to hit the switch of a Formula car's rain light slightly before the normal braking point to make him drop back. When you find that he has made a mistake or dropped back a little for any reason, it is time to focus all of your attention on driving as precisely and cleanly as possible to stretch out that lead.

Here are some basic rules for defending your position:

1. **Always know where your competitor is.**

2. **Pay attention to where he is trying the most seriously to pass.** This is likely to be where his car works better than yours.

3. **Make your car wide.** Don't endanger yourself or your competitor, but make it as difficult as possible for him to put his car beside yours.

4. **Don't neglect your own driving while putting your car on the piece of road your competitor wants.** Drive smoothly and cleanly and use the proper fast line when possible.

5. **When you are able to pull out a small lead, concentrate on putting in a couple of "qualifying laps" to stretch that lead.**

European Attitudes on Racing

We as Americans have developed attitudes on everything from politics to sex that differ significantly from attitudes held in the rest of the world. Perhaps this is due to our isolationist tendencies throughout history or to our rebellious beginnings, perhaps to something else. Whatever the cause, these very different attitudes have influenced the way we race. Oval track racing is the quintessential form of racing in America. Although it has been adopted in Australia and is sometimes practiced in England, oval track racing is very much an American invention and is not widely accepted throughout the rest of the world.

The rolling start is another American invention that, of course, has its roots in oval racing. It is currently used in almost all forms of motorsports in the U.S. Rolling starts seem to be the bane of foreign drivers who come to this country to run Indy Cars or Indy Lights. Many of these drivers have never experienced anything other than a standing start drag race to the first corner in their native lands.

Not only are racing formats different in other parts of the world, but attitudes on how to do road racing differ, too. In a March 1995 article in *Racer* magazine, Jeff Krosnoff, a driver born and raised in California, talks about his experiences running the F/3000 series in Japan, which he did for several years. He explains, "By racing there, I have been exposed to the European attitude of how to go racing—which is much different than the American attitude. This is coupled with the Japanese fastidiousness that means you have to go quick to win, but you can't crash." He continues, "There's a heavy European influence in Japan, so it's not just the Japanese style of driving. You've got a great diversity over there. Here, we don't tend to be as aggressive, but over there I learned to be more aggressive because the Europeans don't take no for an answer. Then again, like I said, the Japanese don't like crashers, so there's not a lot of wild driving going on."

The attitude that Jeff has adopted toward racing since moving to Japan is markedly different from the one he held while in the states. He continues, "If the car is good and you're in a fairly good position on the grid, you've got to be on the attack from the green light; you just have to attack all the time. You try to gain as many positions as you can at the start, but then again, you have to temper that, and avoid over aggressiveness and taking yourself out. When I was racing here, they tended to take the first lap a little easy and then started to race when things got sorted out. In Japan, the instant the green flag falls, you have to start picking people off. You can't be stupid, but if there is an opening there, you have to try it."

If you plan to race overseas at some point in your career, it is imperative to take Jeff's advice and get accustomed to the European attitudes early on. If you race in the United States, by using this European strategy, you can gain an advantage over those who would rather let the field get sorted out before they begin to race. Experienced drivers who have competed overseas may have other bits of wisdom, too, so seek them out and ask their advice whenever possible.

Tragically, Jeff Krosnoff was killed in an Indy Car race in July 1996, a victim of another driver's mistake.

Race Strategy

Concerning race strategy, the late Colin Chapman is credited with saying, "Go like hell from the start and pass as many people as you can." Although this plan works for many drivers in short races, longer races and special circumstances usually require a more sophisticated approach. For any type of race other than a sprint (and sometimes for those, too), you should analyze your strengths and weaknesses and those of your competition. Before the SCCA Runoffs™ at Road Atlanta in 1991, Dave Salls and I came up with the following list of pros and cons for our F Production team versus our serious competition.

	Good Points	Bad Points
1. Dave Salls	Smooth and precise driver, better handling and cornering power, softer tire compound, qualified third.	Deficient in horsepower.
2. Jeff Lane	Excellent horsepower, uses data acquisition, qualified fourth.	Somewhat inconsistent driving may have engine reliability problems.
3. Larry Moulton	Smooth and gutsy driver, reliable car, has won Runoffs™ before, qualified second this year.	Possibly deficient in handling and cornering power, heavy car.
4. Joe Huffaker	Excellent horsepower, well-prepared car, qualified on pole.	Has never run Runoffs™ before, used qualifying tires to get pole.
5. Jack Brock	Good driver, but makes occasional mistakes, good horsepower.	Possible handling problems, we outran him easily earlier this year, qualified eighth.
6. Bob Boig	Good horsepower, consistent driver, good reliability, has won Runoffs™ before.	Car heavy and lacks grip, has not been fast lately, qualified fifth.

All of these cars except Boig's and Moulton's were Midgets. Boig drove a Fiat X-19 and Moulton, a Turner.

From this list, we determined that our most pointed competition would come from Moulton and Lane, but we felt that Lane might not finish. Due to his inexperience at the Runoffs™, Huffaker was not expected to be in the front group at the conclusion of the race. Brock and Boig we considered to be long shots. Since we were starting third, our strategy went like this: We expected a drag race to the first turn because we felt that Huffaker and Lane both had superior horsepower. We did not expect to or care about winning that race. We wanted to avoid a first lap incident at all costs. Races are not won on the first lap, but many have been lost then. After everything got sorted out, we wanted to be in a position to pick them off one at a time by capitalizing on each driver's handicaps and using very clean, precise driving to maximize our speed. Everyone else started on a harder tire compound than we did. If our gamble on tires did not work, we would be slow at the end of the race, but we had every reason to believe it would work. We thought our superior suspension would allow us to run that much softer.

When the race started, it went pretty much according to plan. After an extremely slow start, Huffaker's inexperience with Runoffs™ pressure let Moulton get the lead. Lane won the drag race of row two, and Salls fell in behind him for fourth. By the third lap, Salls had picked off both Lane and Huffaker and was second. He then set his sights on Moulton, who had pulled out a lead of a few car lengths, but he reeled him in and passed him on lap four for the lead. On lap five, Huffaker let Lane and Boig by. On lap six it was Salls by 6/10 of a second, then Moulton, Lane, Boig, Brock, and Huffaker. By lap eight, Salls had stretched his lead to 2 1/2 seconds, but bobbled in Turn Three and allowed the pack to catch him slightly. Boig soon spun and let Brock by, and Lane started loosing a fluid and began to drop back. On lap fourteen, he parked it. On lap sixteen, Salls was approaching lapped traffic, and it slowed him down momentarily. On the eighteenth and final lap, it was Salls, with 7 seconds back to Brock, Moulton, and Huffaker, but Huffaker spun at Turn Five where he had bobbled twice before. He got it pointed in the right direction before he was passed and finished fourth.

Salls won the race, and our strategy and gamble on tires worked, but he had a bit of trouble on the "nineteenth lap," the part of the race that took place in the tech barn. That part of the story is told in agonizing detail in Chapter 4, SCCA 101.

Many factors can conspire to keep you from running a race as you would like to. Regardless of how much analyzing and planning goes into your race strategy, the race will seldom go exactly as planned. Dave and I were lucky that our race went according to our plan that year. With that luck, though, we had an accurate assessment of the strengths and weaknesses of our competitors and ourselves, because we tried to be objective and realistic. You should remember that each of your competitors will have a strategy, too, and theirs may conflict with yours. If a competitor views your strength as having better horsepower than he does, he may try to get a better start and win the "drag race" against the odds. If he succeeds, he may then be in your way when you try to use your strategy to use your horsepower to get by a car in the row in front of you. Plan your strategy as carefully and as accurately as you can, but be prepared to throw it all out the window the second things deviate from your plan.

In addition to the priorities involved in driving, i.e., controlling line, regulating speed, and maximizing traction, and the two involved in racing which have already been discussed, passing and pass avoidance, two others priorities should be set during the race: managing tires and judging traffic. If your racing includes pit stops, chassis setup can be added to the list, but we'll deal with that in the next chapter.

Many club racers try to maximize tire life by using the same set for three or more race weekends. If you are consistently faster than your competition even with used tires, this may be acceptable. Tires lose their tractive capacity as a function of how much time the plasticizers in the rubber have to evaporate. This happens much quicker at high temperatures than it does at room temperature. The plasticizers in a tire subjected to sustained high temperatures will evaporate very quickly, but the rubber on the tread of the tire is being worn down quickly as well. Indy and Winston Cup cars only heat cycle their tires once, but abrade away a lot of rubber in the process. Club racers don't chew the rubber off of the tires as quickly, but the time between heat cycles is death to the rubber compound. Once a tire is scrubbed in, it quickly loses its tractive capability, and within six weeks it will be a couple of seconds slower even if it has only been run once.

For most other racing, tire life is measured in "heat cycles." This term refers to the number of times a tire has been heated to racing temperatures and allowed to cool. If this happens four or five times in a weekend, which is typical of a Regional/National event, the evaporation continues to take place when the tire cools, but at a reduced rate. It even continues between events. By the time the third event is run, it is likely that the tractive capacity of the tire has dropped off sharply even though a large amount of tread rubber is left.

It is also possible to overheat a tire only once and have its performance fall off. For this reason, if your race is a long one, or if you suspect the tires may go off before the end of the race, care must be used in driving to maximize tire life. Soft tires will generally stick better, all things being equal, but will have a shorter useful life. To properly manage tire life in a race, the driver should always be aware of the level of abuse that he is subjecting his tires to. Wheelspin, locking tires, and sliding the car are murder to a racing tire. If the car is not properly set up, some sliding may be required to get the car around the track, but this always results in shorter tire life. Sliding or spinning a tire not only abrades away tread rubber, but also elevates the temperature of the tire. If the car is set up well enough to allow the driver to put in fast laps without unduly sliding the car, tire life will be substantially longer and/or a softer compound may be used.

Working traffic well may mean the difference in many positions at the end of a race. If you are involved in a battle with one or more cars, traffic can get in the way when lapping other cars or when other competitors reenter the race after pit stops. If these competitors are not racing with you for position, racing etiquette says that they should move over and let you by. In practice, though, this does not always happen. A driver may try to get out of your way, but move the wrong way. He may see your competitor, but not you, or he may simply be involved in a battle of his own and not want to give his competitor an advantage by moving over for you. Every situation is different, but a driver should develop the ability to recognize traffic patterns and analyze how he might be able to use certain situations to pass his competitor or put a slower car behind him and between himself and his competitor. Study these situations when they arise on the track. After the race, analyze them to learn what you or your competitor did correctly and how you could use the situation better in the future. Watching races on TV is a good way to study traffic situations, too, especially since on-board cameras have become so good. The ability to work traffic effectively is the mark of a seasoned driver.

When races include planned pit stops, things get a bit more complicated. On the surface, a one-pit stop race is relatively easy. You start the race with just a little more fuel than necessary to get to the halfway point of the race, refuel, take on new tires, and go to the finish. In practice or testing for a given race, however, many teams analyze the performance of their cars with decreasing fuel load and decreasing tire performance. Each of these factors will have a significant effect on lap times over the course of several laps. Race engineers analyze lap times to determine exactly how much lap time increase can be expected per lap due to tire wear and how much the lighter fuel load will decrease times per lap. For a one-stop race, it may be possible to stretch the first on-track session of the race to the limit of fuel capacity or tire performance before the stop and pick up the race afterward with a lighter fuel load than the competition and fresher rubber. If too much time has not been lost to the slippery tires prior to the stop, this may give the team a substantial advantage toward the end of the race. A great deal of thought goes into the pit stop strategy for a race to secure an advantage. With two or three stops, strategy can become very complicated.

In the event of a nonscheduled stop or a deviation in plan due to an opportune yellow flag, a contingency plan must be enacted. Some basic plans may be pulled off the shelf on these occasions, but conditions cannot be accurately predicted, so some improvisation is then required. When race engineers are shown on TV studying laptop computer screens during a race, they are sometimes trying to reevaluate their strategy after something has gone wrong. To be successful, racers must roll with the punches and make the best of imperfect situations.

Chapter 8

Chassis Setup

You have to know how the chassis works and how the tires work. You have to know everything if you are going to run a car at its limit.—A.J. Foyt

Chassis setup is the single most important factor in making a race car fast at any given track. It is more important than having the perfect ratios in the gearbox or having a well-tuned engine that makes a great deal of torque and horsepower, although each of these is important too. It may sound like an oversimplification to say that the advantage of a well-set up car is that it allows the driver to keep the throttle open longer and use more of the engine's power, but it is true. Attaining the proper balance of aerodynamic drag and downforce is also part of set up, as is choosing the proper tire and/or tire compound, and all of these must be optimized to suit the driver's style and the prevailing track and weather conditions. Chassis setup is an extremely complicated facet of winning races, and is often underestimated by the amateur racer.

Individuals who become really good at chassis setup can, eventually, call themselves racing engineers. Those who excel may become car designers. But it is a very rare driver who is good at both driving and engineering. For this reason, in the modern world of motorsports most successful drivers choose to work with an engineer. A driver usually has all that he can handle in just driving the car. He needs to come off the track, be debriefed by the engineer concerning the car's behavior, perhaps participate in examining graphs from the data system, and then go off to relax. The decisions concerning what should be done to the car to improve the setup should be made by the engineer, giving the driver time to wind down and prepare for his next on-track session. If the driver is a competent engineer, he should be included in deciding which changes should be made to the chassis or at least given the opportunity to agree with the engineer's assessment. *Then* he should relax.

This is a very important point which is often neglected by club racers. No part of the car/driver combination is more responsive to proper tuning than the driver himself. His job is to drive the car as fast as he is able. His abilities will always be greater if he is rested, alert, happy, confident, and relaxed. At the track, the driver should not work on the car!

This does not relieve the driver of the responsibility of having some knowledge of what is involved in setting up the car, for a number of reasons. First, in the early stages of his career, finding a proficient engineer to work with him (for free) may be almost impossible. At this time, it is especially important for the driver to understand how to make a car handle. An ill-handling machine has stopped many driving careers prematurely. At best, a poorly set up car will add to the frustration level and slow the driver's progress considerably. Second, since an incorrect setup will slow the car, the driver's finishing positions will be lower than what should be expected. Third, and more important in the long run, the driver must develop very special relationship with his engineer, one that is based on technical knowledge of the chassis and a common vocabulary of technical terms used in their conversations. It is essential, then, for the driver to be fluent in the concepts of vehicle dynamics and how those concepts may be implemented on the car he is driving. This chapter is essentially a crash course in vehicle dynamics. It is not intended to be a substitute for the proper study of the subject which is necessary for a more complete understanding. See Appendix B for more sources of information.

Where the Rubber Meets the Road

> *Whenever weight is transferred between a pair of tires, the result is a net loss of traction.*—First Law of Vehicle Dynamics

A tire is composed of two basic elements: the carcass and the rubber that encases it. The carcass consists of two layers of a synthetic fabric (usually nylon) which are oriented on an angle to the longitudinal axis of the tire. This angle is called the bias angle. In the case of a radial ply tire, the bias angle is 90 degrees meaning that the threads of the fabric run with and at right angles to the direction of motion. The bias angle is important because it determines the resistance of the tire to being twisted by the steering wheel or other forces about the area that is in contact with the pavement. The area in contact with the pavement is called the contact patch.

Rubber in its natural state is light brown, almost amber in color. It does possess some elasticity, but could hardly be called durable. Natural rubber is much like an artist's gum eraser. Carbon is added to the rubber to give it strength and toughness. Polymers and oils are compounded with it to further alter its characteristics. The proportions of these basic ingredients determine how the rubber in the tire stretches, how adhesive it is when hot, and how resistant it is to abrasion.

Rubber compounding and carcass construction are regarded by most outside of Akron as "black arts." Tire engineers rarely let this sort of proprietary information get out, so the myths are perpetuated. In reality, though, there are only a few basic guidelines for formulating tires to be stickier, stretchier, or longer-lasting. It is not necessary for those of us who use tires to become tire engineers, since we cannot change what we get from Goodyear, Hoosier, Bridgestone,

Yokahama, and the rest. However, since the tires on a car have the most direct influence on its handling, we do need some understanding of the conditions for which the tires have been designed. We can then ensure that those conditions are present to optimize the tires' performance.

Heat is generated in a tire from the friction as the tire slides or spins, even a little, in or away from a turn. Even more heat is produced by stretch of the rubber as the sidewall deflects and the tread is twisted around the contact patch. The rubber in the tread of a racing tire is intended to operate within a certain temperature range. Using the tire above or below that optimum range spells trouble. If too much heat builds up inside the rubber of the tire and it overheats, the first symptom is the tire getting slippery. If the tire continues to be operated above the maximum allowable temperature, it will soon develop blisters where small patches of rubber separate from the carcass. This is shortly followed by large chunks of tread flying off the tire. Drivers have seldom reported this experience to be pleasant.

Overheating of a tire is most often a result of a combination of rubber too soft for the weather or weight of the car, too much camber or pressure, and a driver who throws the car around too much or a chassis set up with a lot of oversteer or understeer. Because of these factors, a tire will usually be overheated either in the center if too much pressure is used, or on the inside edge if the wheel has too much camber. It is possible, although not as frequent, to overheat the entire surface of the tire. Regardless of what part of the tire is overheated, once it has happened, the tire is junk. Even if no obvious damage has occurred, it will never again be fast.

Underheating a tire is not quite as serious. No detrimental effects (usually) show up as a result of running a tire too cool, except that the car is slow. If the temperature of the tire is not up to the proper range, it is because the rubber compound is too hard for the weather or car weight, the tire is old and the rubber has hardened, or the driver is not pushing hard enough. Occasionally, however, tires running too cool can cause a major problem. You may remember the 1992 Indy 500 race day which turned out decidedly cool and surprised the Goodyear engineers. Having only harder rubber suitable for a warmer day, many cars were written off as soon as the green light was shown on restarts. The tires that were running too cool to start with cooled even more during yellow flags, and were stone cold when the green again came out. This scenario was played out over and over again during the race. Another situation that has surprised some Indy Car drivers is getting fresh rubber during a stop, lighting up the rear tires when leaving pit lane to get them hot, reaching the first turn, and finding that the fronts are still cold. This usually results in the car understeering into the wall and a short race.

The operating temperature of most racing tires should be between 190 and 220 degrees Fahrenheit. Tire-makers produce tires with a variety of rubber compounds for each tire size, so that they can be selected to stay within this temperature range when conditions vary. Goodyear Formula Ford tires, for example, are offered in 100, 160, 240, and 430 compounds. (The lower the number, the softer the rubber.) Fords, being quite light, can use soft tires, especially in October at Mid Ohio. Trans Am and GT-1 cars, on the other hand, are comparatively heavy and produce a great deal of torque, enabling them to light up the rears easily exiting slow turns. For this reason, they typically use a 600 compound, but more adventuresome drivers sometimes try the softer 430 compound. This compound in Formula Ford tires lasts a long time, but Ford tires that are this hard only get up to temperature in Texas in August.

When the compound is right for the prevailing conditions, the rubber actually becomes adhesive. It is like running tires made of tape, sticky side out. Obviously, this uses a lot of power down the straight, but speeds can be quite high through the turns when the tires are working. And, as we saw in the Chapters 6 and 7, one of the driver's major concerns is to exit the turns as fast as possible.

When a tire is "glued to the track" in this fashion both by its adhesive qualities and by the weight resting on that tire, the contact patch tends to remain in the direction the tire is traveling when the steering wheel is turned at the entry to a turn. The wheel, along with the sidewall on the tire, changes direction as the steering wheel is turned, but the contact patch, for a short time, remains oriented straight ahead. This difference in angle between the direction of the wheel and sidewall and that of the contact patch is referred to as the slip angle. Whoever coined this term, though, should be made to go around the world's paddocks explaining it because, in reality, no slipping is occurring. Rather, the elasticity of the tire pulls the contact patch into line with the new direction of the wheel, and this causes the car to turn. If the car used tires with very little elasticity, steel tires for instance, it is intuitively obvious that the car would not turn. The elasticity of the tire makes the car change direction. The force with which the contact patch is stuck to the pavement and the "stretchiness" of the rubber/carcass combination produce the forces of lateral acceleration.

The concept of slip angle is an important one. A tire's slip angle characteristics, given a certain amount of friction with the road surface and weight placed on the tire, will determine how it responds and how much room the driver has with which to work to keep the tire at its maximum efficiency. Figure 8-1 shows the relationship of lateral force to slip angle for two typical tires. In each case, the slip angle curve rises rather linearly until the elasticity of the tire can no longer accommodate more force, at which point the curve begins to level off and then decline. The point at which the curve begins decline is the point at which the contact patch can no longer stay glued to the road and "sliding" begins to occur. The driver will feel this reduction in contact patch area on the front as a lightness in the steering followed by understeer, and on the rear, as the car getting "loose." The more gradual the change of direction at the top of the slip angle curve, the broader the tightrope on which the driver can balance his car. Radial tires have sharper curves than bias ply tires and therefore require more sensitivity on the part of the driver to keep from falling off of it, but the curves generally peak at higher values.

The weight of one corner of the car that is supported by a tire is called, in engineering terms, the normal force. Normal refers to a force oriented 90 degrees to the resisting plane, in this case, the road surface. Isaac Newton showed us that the lateral force required to move any given object rises in proportion to the normal force applied to it. Place this book on a table and a certain force will have to be applied to the edge of it to make it slide. Place another copy of it on top (you did buy two, didn't you?) and it will take twice as much lateral force to make the stack of two books slide.

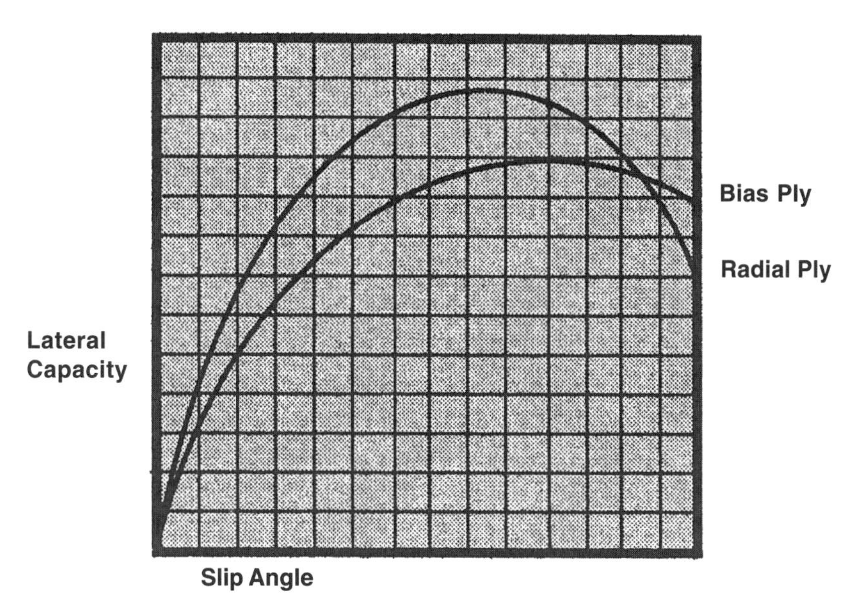

Fig. 8-1 Radial tires can generate higher cornering forces, but over a narrower slip angle range than bias tires

Newton's observation holds true for inelastic bodies (no body is completely rigid, but that is a completely different subject), but falls a little short where elastic tires are concerned. This relationship of normal to lateral force holds true up to the limit of elasticity of the body, but the more elastic the body is, the more gradual the breakaway will be, hence the curve at the top of the graph shown in Figure 8-2. Since both the lateral and normal forces are measured in pounds, dividing one by the other cancels the units and leaves a dimensionless number called a coefficient, in this case the coefficient of friction. The coefficient of friction (C_f) rises with increasing normal load, but begins to fall off in the upper ranges of increasing load. It can be seen that the capacity of a tire to handle an increasing lateral load can be increased by increasing the normal load applied to it within the linear range. This is why we put wings on race cars.

Since we want to operate a tire with as much normal load as possible to maximize its lateral load-carrying capacity, we usually operate it pretty high on the coefficient curve. The difficulty arises when the car changes direction, and the centrifugal forces act to pull the tire toward the outside of the turn. The response to this centrifugal acceleration is weight transfer from the inside tires to those on the outside, causing them to operate even higher on the coefficient curve, where increasing vertical load produces proportionately less lateral capability. The inside tires, meanwhile, are operating lower on the coefficient curve, where they provide less lateral capacity than they did before the change of direction was initiated. The result is that the inner tires provide less grip than before, and the outer tires provide more, but not as much more as was lost by the inner tires. This is an important concept since all of chassis setup is based on it. Refer again to Figure 8-2 to see how it works. This is the basis of the First Law of Vehicle Dynamics: *Anytime weight is transferred between a pair of tires, a net loss of traction results.* If you never remember anything else about vehicle dynamics, remember this.

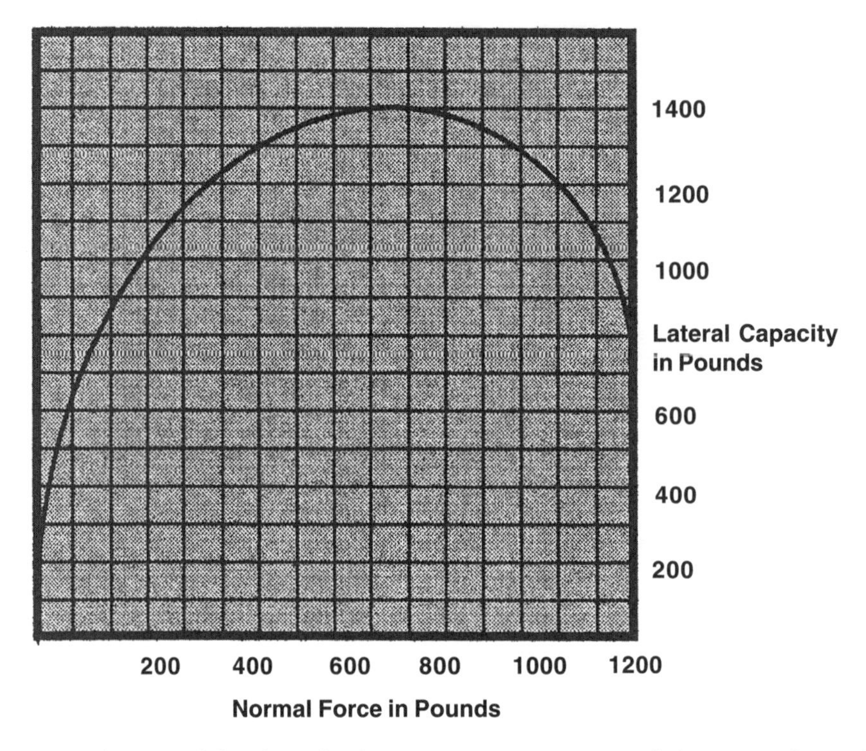

Fig. 8-2 Increased vertical load applied to a tire increases its ability to withstand lateral force—up to a point. Dividing lateral capacity by normal force at any point gives a tire's friction coefficient.

Weight Transfer

It should now be apparent that any car will be faster through the turns if it transfers less weight laterally. Methods to reduce this weight transfer are easy to explain, but somewhat more difficult to implement. The two most basic approaches are to lower the mass of the car and to make the car wider. The relationship between these two approaches is shown in Figure 8-3. If a triangle is drawn between the center of gravity, a point on the track surface directly below it, and the center of the tire contact patch for two different cars, the car with the smaller angle between the track surface and the hypotenuse (the slanted side of the triangle) will transfer less weight. A major reduction in weight transfer requires that the center of gravity be lowered or the track width be increased.

Using either of these methods to increase the cornering capability is most easily accomplished in the design of the car, but it can sometimes be accomplished by small changes in wheel offset, A-arm length, or other modification. Some years ago, a significant wheelspin problem on an F/2000 car that I was working with was completely eliminated by increasing the track width by only one inch per side. This modification kept more weight on the lightly loaded inside rear tire, which is prone to spin on an open differential car. Always keep the components of a car as low as possible and use the maximum allowable track width.

Weight transfer is most profoundly affected by these two factors, but some minor factors have an effect, too. Of these, the most important is the relationship of the height of a suspension's roll center to its track width. The roll center is an intangible but important point about which one end of the car rolls as weight is transferred to the outside wheel in a turn. It is defined by the suspension linkage, and is the point of the resolution of forces that are applied to those links when cornering. The higher this point is above ground level, the more weight is applied to the outside wheel. The roll center also acts to apply an upward force to that end of the chassis, which is the notorious "jacking force." Roll centers should be as close to the track surface as possible to minimize the weight transfer. However, their exact placement is dictated by the amount and direction of their movement during chassis roll and pitch, as well as desired camber curves, track width change, and a host of other parameters. The design or modification of suspensions is an extremely complex area, and may require outside help unless your team's engineer is very well versed in the subject.

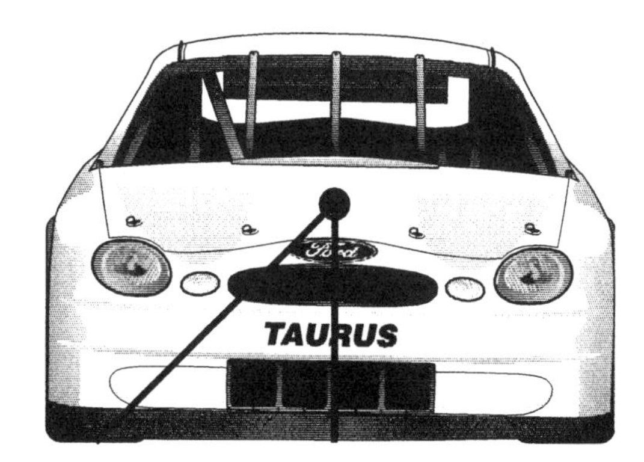

Fig. 8-3 The car with the lower angle between hypotenuse of the triangle and the track surface will transfer less weight laterally when cornering. In this context CG height and track width are synonymous.

Aside from working on the car to reduce weight transfer, smoothness is also an important factor, and the reason that books and driver's schools stress smoothness should now be evident. Tires do not respond favorably to rapid increases or decreases in the weight they support. Such rapid changes cause the ratio of vertical to lateral load to change rapidly, which frequently overstresses the elastic capability of the tires and causes the contact patches to break away from the track surface prematurely, before the maximum coefficient has been reached. To be fast, decrease the amount of weight that is transferred and lower the rate at which it is transferred by driving smoothly.

Steady State—Balance

Weight transfer and tire selection mainly affect the generation of cornering force. Most of the remaining several dozen chassis-tuning factors affect the car's handling. These are two completely different areas, but when each is properly tuned, they should be complimentary. It is possible for a car to generate a great deal of cornering power, but be so crippled by poor handling that the cornering power cannot be used.

A few years ago, I conducted a series of seminars on vehicle dynamics to teach club racers how to set up their cars to make them faster. At one of these seminars, a student continually asked questions to which he seemed to expect simple answers. However, to properly answer his questions required more complex answers than he was prepared to accept. Finally, I realized that he was looking for the "magic screw" on a car, and wanted me to tell him how much to turn it and in which direction. In reality, there is not only one; there are several dozen magic screws all over the car, and each one of them must be turned the proper amount in the right direction *in conjunction with all of the others* for the car to be really fast. A lot of effort must be expended on chassis setup to get close to the optimum setup and to get the last little bit of speed out of a car.

Handling is a catchall phrase which is more specifically defined as the combination of balance, stability, and response. Let's deal with balance first. When the car is balanced, it is said to be neutral. On either side of this neutral condition are oversteer and understeer. Briefly, oversteer occurs when the car seems to rotate with the rear moving toward the outside of the turn, resulting in the front of the car pointing into the turn more than the steering angle indicates. When the front of the car does not seem to respond to steering input, but lags behind the direction in which the front wheels are pointing, the car is said to be understeering. Intuition tells us that a neutral car would be the fastest of these three. Things are not always as they seem, though, and we find setups on either side of the neutral condition to work better under certain circumstances.

In 1994 and 1995, Tony Cicale was Jacques Villeneuve's engineer at Team Green. He has had a long and quite successful career, including building his own 2 liter sports racing car in the '80s and designing the aerodynamics for the VDS Can-Am cars, in which Geoff Brabham and Michael Roe each won championships. Tony explains proper balance this way: "Overall, the optimum setup in terms of performance is probably slightly loose, which is how a lot of cars are set up for qualifying. In terms of having a consistent racecar, with which your driver will consistently do better lap times, a slight push is an optimum setup for the start of the race."

The driver's style must be considered in deciding on the proper setup for a car. A driver who is smooth and sensitive can more easily stay on top of a slightly oversteering car. A driver who throws the car around at the drop of a helmet might be more apt to make a mistake unless his car is set up to understeer slightly. As a matter of fact, I once had a driver get out of a car and pronounce it to be oversteering significantly after another driver had just said that the same car understeered badly. We must not regard the car too objectively as having inherent characteristics apart from those of the driver. It is the car/driver system we are trying to tune to be fast.

Oversteer and understeer can each be induced artificially. Lifting the throttle foot for a moment while cornering will bring the rear out. In a high-powered car, lighting up the rear tires at the exit of a slow corner will do the same. Conversely, when the brakes are still on hard when the car is turned into a corner, understeer will result. These responses should not be confused with the basic tendencies of the vehicle. The basic tendencies are most easily seen in steady state cornering.

The two Club Fords in the front on this oval appear to be balanced quite neutrally. Number 36, however, is understeering or "pushing," causing him to take a lower line into the turn to avoid a meeting with the wall at the exit. This is hard on front tires and results in higher lap times, but, depending on the driver's style behind the wheel, his sensitivity, and the length of the race, it may be a better setup.

Few corners on road courses and street circuits allow a driver to experience steady state cornering. The ones that do are long, constant-radius turns. The "carousels" at both Road America and Mid Ohio are two of the best examples. Some ovals also include constant-radius turns. Turns Three and Four at Phoenix are such turns, although Turn One is sharper and opens into Turn Two. Skid pads, being circular test tracks, provide the best places to establish steady state balance.

It is fairly easy to set up a car to do what you want it to do in a steady state. This usually requires working with the spring and anti roll bar rates. Although some engineers use a specific formula relating spring rates to unsprung weight and dynamic weight, the more common method of determining spring rates is to soften them until too much dive is experienced when braking and too much squat is generated when accelerating in the low gears. In the absence of more scientific methods, this is a good place to start. If the car bottoms too easily at a normal ride height, the springs are probably too soft. (Or perhaps the track is rough, requiring a higher ride height. It's never simple.)

In addition to keeping the car from bottoming, stiffening the springs will make the car more responsive. This can be either good or bad. Softer springs give an inexperienced driver more time to react to the car's responses to the forces acting on it. Stiffer springs allow less time for those reactions, but make the car respond more quickly to the driver's control inputs. Generally,

Making a spring rate change will alter the response characteristics of the car in both roll and pitch. Generally, soft springs allow the car to roll and pitch more and give the driver more time to react. Stiffer springs reduce roll and pitch (and tire compliance), but make the car more responsive to steering, brakes, and throttle.

the more experienced the driver, the stiffer he prefers his car. Some top drivers set their cars up like boxcars! Many more factors influence spring selection, such as the traction characteristics of the tire/track system, the bumpiness of the track, the allowable range of ride heights, and—according to Carroll Smith—the phase of the moon. The object of selecting springs is to keep the tires in constant contact with the irregular surface of the track while limiting the pitch motions of the chassis. Spring rates are rarely ever perfect.

While the springs have an effect on the pitch motions of the car, the anti-roll bars affect the roll motions. It is important to keep in mind that these motions are caused by the weight transfer due to the centrifugal and longitudinal forces acting on the car's mass, which is above ground level and above the contact patches where the forces originate. The springs and bars do not cause the weight transfer, and changing their rates does not change the total amount of weight that is transferred. Anti-roll bars control the response of the chassis to the weight transfer caused by the centrifugal force. That centrifugal force is constant for a specific car on a turn of a given radius taken at a given speed.

An anti-roll bar connects the wheels of each end of the car, and causes an independent suspension to be a bit less independent. As you will remember, the transfer of weight from the inside to the outside tire in a turn decreases tire performance. In this case, however, the bar performs two functions which, taken together, more than make up for its detrimental effects. First, it keeps the door handles from rubbing the ground in high G turns without using springs too stiff in vertical rate. Additionally, the ratio of the weight that is transferred at each end of the car can be altered by changing the relative stiffnesses of the bars. The total weight transfer cannot be changed, but its front to rear proportions can be. Changing the dynamic weight on each tire in a corner, as we have seen, alters the point on the slip angle curve at which the tire is operating. In this way, an understeering car can be made to oversteer by softening the front bar or stiffening the rear. An oversteering car can be made to understeer by doing the opposite.

When the car rolls to the outside of a turn, the anti-roll bar is twisted. The stiffer the bar, the more the inside tire is unloaded, and the less total traction is available. This detrimental factor is minor, though, compared to the ability to alter the dynamic weight distribution between the front and rear.

Along with the bars, the springs contribute to a car's roll stiffness. Some reasonably well-balanced cars have used very stiff springs and puny bars and exhibited no more chassis roll angle than a more conventional car with softer springs and stiffer bars. We must question the tire compliance of the former group, though. Once the spring rates are determined, the total rates of the two anti-roll bars should keep the chassis roll angle down to manageable terms. Giving the corner workers a look at the underside of your car may make interesting conversation at the Saturday night party, but it is not often the quick way around. Letting the car roll excessively uses bump travel of the outboard wheels and makes it difficult to control camber and roll center movement.

Camber is the angular relationship between a wheel's vertical center line and the road surface. Some books on vehicle dynamics relate camber angle to the horizontal plane of the chassis, which is only important when the chassis is parallel to the ground, a rather rare dynamic occurrence. This reference is often used, though, and we will also use it at some points in this discussion. Be sure to always know if the camber angle being discussed is relative to the chassis or the track surface. The angle between the wheel and the road is important in maximizing the amount of rubber in contact with the surface, and also to distribute the loads as evenly as possible across the face of the tire. Just as traction is lost when weight is transferred laterally between a pair of tires, a single tire suffers a loss of traction when the load is not distributed evenly across it. The goal is to keep the tire flat on the track surface at all times. Due to the pitch and roll motions of the car in response to the various accelerations acting on it, this goal is not an easy one to achieve.

Camber thrust is also important. When a tire is operated with dynamic negative camber in relation to the track surface, it is more resistant to sliding toward the outside of the turn. For a demonstration, hold a pencil with the rubber eraser down at a right angle to a tabletop and slide the eraser across the table. Now lean the pencil so that the eraser is leading the pencil and repeat the movement. At this angle (analogous to negative camber), the inside edge of the eraser tends to fold under, raising the pencil slightly and resisting the sliding motion. On a race car, the amount of negative camber used must be tempered by the need to properly distribute the forces across the face of the tire. Finding the proper balance point between these two is difficult, especially since it is not the same for tires of different ply angles. As a general rule, radial tires like more camber than bias tires.

With a double A-arm suspension, the upper arm is always shorter than the lower one to tip the top of the wheel inward as it moves up into bump travel, which increases camber in the negative direction. Depending on the relative lengths and angles of the arms, the opposite is usually true as the wheel on the opposite side moves down into rebound travel. The camber angle of the inside tire in rebound is of little significance, however, when most of its weight has been transferred to the outside.

The generation of negative camber is a function of the design of the suspension. The camber curve of a given suspension, then, is set from the beginning and cannot be changed. The point on the camber curve at which the car is operated can be changed, though. Altering the car's ride height changes the portion of the camber curve which is used. Anytime the ride height is changed, the static camber angle must be readjusted. The varying camber angle (with respect to the chassis) as the wheel moves into bump travel is essential in order to keep the camber angle in relation to the track zero as the chassis rolls in a turn. Ideally, the static camber should be zero, and the increasing negative camber with increasing bump travel should equal the chassis roll angle at any point. In this way, the tire is flat on the road at all times.

In practice, other design considerations, particularly the roll center position and its movement, keep this goal from being realized. Rarely ever will an independent suspension produce enough negative camber change. This is the reason that some camber must be set in the suspension statically. The correct amount of static camber is determined by measuring the temperature of the tire tread at three points, the center and each edge, as soon as the car comes off the track. Not enough negative camber will show up as an outside edge hotter than the other two points. Too much negative camber makes the inside edge too hot. The more sophisticated and sensitive the suspension (and the less wheel travel used), the less static camber is needed to keep the tire temps equal.

Other types of suspensions each have their own problems with camber. Explaining in detail how to deal with each of them deserves a volume of its own, so I will not go into this here. For more information, refer to the works listed in Appendix B dealing with Vehicle Dynamics.

Aerodynamics also plays a part in a car's steady state behavior, especially on cars that use wings or ground effect, but only at higher speeds. It is best to balance the car at slower speeds first through mechanical grip with the springs and bars, and then adjust the aerodynamic aids to give it the same balance characteristics at higher speeds. These aids have little effect at low speeds.

Formula One, Indy, and Atlantic cars that use ground effect tunnels receive the major portion of their downforce from these devices, and wings are used for fine tuning high-speed balance. Cars that have no tunnels, such as F/2000s, make much less downforce, but still are balanced by adjusting front and rear wing angles. Cars that use neither wings nor tunnels rarely have air dams and spoilers altered to adjust high-speed handling, because these items make relatively little difference, and because in setting up these cars we are looking for all of the downforce (or reduction of lift) we can get. Spoilers and air dams are usually of maximum allowable size, and are not adjusted so as not to lose any of the precious little effect they do produce.

We have just used fourteen "magic screws" to balance the car for steady state behavior: four springs, two bars, four camber settings, ride height at each end of the car, and two wings. Steady state cornering should now be acceptably close to neutral. But transitional balance is encountered at the entrance and exit of each steady state turn, and also comprises the major portion of each non-steady state turn. So, even though steady state balance is important in the general scheme of things, the transitional characteristics of stability and response are even more important. A car contains many more magic screws affecting transitional balance than those affecting steady state balance, so things start to get rather more complicated here.

Transition—Stability and Response

As discussed in Chapter 6, each turn can be thought of as having three phases: the entry, the mid part, and the exit. The entry is defined as the area from the turn in point to the point of throttle application; the mid part, from throttle application to full throttle hard acceleration; and the exit from full throttle to the track out point. The balance of the car between understeer and oversteer in steady state cornering shows up in the mid part of the turn. Unfortunately, in most turns the entry and exit phases are of considerably longer duration than the mid part. These phases are dominated by the car's transitional handling—the areas where it is changing from one steady state to another. Many factors work together to make a car manageable in these transitional areas. If a car exhibits a problem, it is essential to determine exactly where in the turn the problem occurs.

Generally, the problems that arise are either understeer or oversteer at the entry or exit. It is possible, even common, for a car to change from corner entry understeer to corner exit oversteer. This may be the result of misadjustment for the conditions of any number of elements of the car. To show what is possible, let's begin with an example car showing exactly these symptoms. If the car has a particular resistance to turning in, it is possible that too much front brake bias is being used. The rear brakes do not do enough work, and the fronts do too much, limiting the tractive capacity they have available for turning. Corner entry understeer from this cause should diminish as the brakes are released in the entry phase. A soft front spring and/or shock compression rate may also cause this condition by allowing too much dive of the front end, causing the turn to be initiated on the inside edges of the tires. If it does not diminish but continues through the corner entry, the shock rebound rate may be too high as well, causing the front of the car to "stick" down. Many times a car that understeers through the entry phase will continue to do so through the mid part and into the exit, so it is important to stop it early. (Actually every car has at least a slight amount of corner entry understeer, although on a good car it is usually imperceptible. This happens because it takes a finite amount of time for the elastic deformation of the tire to occur and begin to generate some slip angle after the wheel has been turned to initiate a change of direction.) Let's assume that our example car only pushes early in the corner and then switches to oversteer at the exit.

If it is a slow corner and/or a powerful car, wheelspin is the likely culprit, but this itself may be a symptom of another malady. The oversteer might be traced to a malfunctioning differential, tires too hard or worn, or rear shocks that cannot cope with a bumpy track. If the car in question has very wide rear tires and independent suspension, a small change in camber produced by the rear of the car dropping due to the hard acceleration of a big engine will drastically reduce the size of the contact patches. The probable result will be both wheelspin and lack of lateral grip. This may even be true of lower-powered cars with narrower tires if the rear spring and/or shock compression rate is too soft. If the spring rates have been carefully worked out and this is still a problem, "bump rubbers" may be used on the shock absorber shafts to provide additional compression resistance toward the end of the stroke. These compressible cones are sort of a poor man's progressive spring, but are light, cheap, easy to work with, and effective. Even the high-tech, high downforce cars with their very stiff springs can have this problem. A casual observation of Formula One cars exiting a slow turn at a recent race revealed a visually perceptible and significant amount of rear end squat on acceleration and some occasional oversteer. Of course, another potential cause of corner exit oversteer is a driver who has not learned to modulate the throttle. (Traction control is no longer allowed in F/1.) If this is suspected, however, it does not mean that a higher corner exit speed cannot be obtained by improving the accelerative capacity of the car, too. The driver may be having trouble with his right foot because the tires are gone.

It should be apparent that the reactions of the car in each phase of a turn can take place in different areas of each phase. Corner entry understeer can terminate as soon as the brakes are released, or it can continue deeper into the entry of the turn. To properly diagnose the cause, the sequence of handling events and their relationship to the driver's control inputs and the external factors affecting the car must be known.

In the example above, shock absorbers were mentioned several times as possible trouble spots. These days, the shock absorbers are the most important devices used on a car to change its transitional characteristics. But, as with slip angle, the phrase "shock absorbers" has caused some problems in understanding what these devices do. The British call them "spring dampers," which is slightly more appropriate, but still leaves some gaps in meaning. The current generation of racing shocks do dampen oscillations of the springs, but more importantly, they control the high-frequency motions of the wheel over irregular track profiles to improve tire compliance with the track surface. They control the rate of the chassis's response to the weight transfer caused by lateral and longitudinal accelerations.

All shocks are hydraulic, even the ones that are advertised as "gas shocks." A series of valves with different orifice sizes are incorporated which open under different velocities of the oil flowing through them. Some valves operate at low velocities, some at higher ones. On proper racing

Modern shocks such as this triple adjustable Penske are the most important tuning aids for the transitional areas of a track. They not only dampen the oscillations of the spring, but improve tire compliance over bumps and rough pavement and control the rate of chassis roll and pitch.

shocks, the opening points of the valves are adjustable in order to control the damping of the tire as it rolls over small road irregularities at high speeds, and to control the rate of chassis roll or pitch of much higher displacement, but lower piston velocities. These are adjustable in each direction of the shock piston independently of the other. Koni was once the class of the field in adjustable shocks, but since their time Fox, Penske, and now Q Damper have each made major advances. The new state of the art is the six-way adjustable shock from Q Damper which allows bump and rebound adjustments on low-, medium-, and high-speed piston velocities, with all adjustments being made on a remote mounted reservoir.

Differential Types

A differential is required, theoretically, because the inner and outer driving wheels are traveling at different speeds as a car negotiates a turn. (The outer wheel is farther from the turn center than the inner one and must travel a greater distance in the same amount of time.) Without this differential action, the inside tire would skid across the road surface, as anyone who has pushed a drag racing car around the paddock well knows. On your street car, a differential reduces tire wear in parking lots and generally makes low-speed driving more pleasant. Wheelspin exiting a turn is murder to acceleration, can cause corner exit oversteer, and uses up rear tires prematurely, as well. The "open differential" is very bad in this respect. As the normal load is reduced on the inside tire, the open differential directs more of the engine's torque to that tire, resulting in more power on a tire with less traction. Inevitably, it begins to spin. This forces the car to coast out of the turn regardless of throttle position until traction is regained. In an effort to eliminate this problem, other types of "differentials" have been developed over the years, each having its own best applications.

Many Indy Car race engineers use an open differential at the high-speed ovals because they say it "frees up the car." Translated, this means that since the speeds are quite high and wheelspin is not a problem, the open diff is effective because it wastes less power. It also means they can miss the proper tire stagger, the relative circumference of the rear tires, slightly and get away with it. On shorter ovals, racers sometimes prefer a completely locked differential called a "spool." This type of differential locks both wheels together all the time so that no differential action is allowed. To use a spool, the stagger must be perfect. The car will pull to one side when traveling straight ahead with this arrangement, and on longer tracks the drag of the inside tire is both significant and detrimental. When used on slower tracks, or even in slow turns on faster road courses, a spool can cause corner exit understeer. When this is a problem, some racers use a lot of low-speed rear shock rebound stiffness or a really stiff rear anti roll bar. These can be used either singly or in combination to unload the inner rear tire. This seems a bit like cranial amputation to cure a headache, though. A more appropriate solution may be more rubber up front or a wider front track. When these options are available, a spool can be a good choice for lower-powered cars on relatively fast road courses.

To arrive at a compromise between the open differential and the spool, limited slip differentials have been developed which are completely unlocked when the power is off and lock the wheels together in varying amounts when power is applied. Several types have been produced, including the Cam and Pawl, the Salisbury, and the Detroit Locker. The Cam and Pawl works quite well, but is expensive, non-adjustable, and requires a lot of maintenance. GM's famous "Posi-trac" is a clutch plate type Salisbury limited slip. The clamping force of the clutch plates can be adjusted when the unit is set up to determine how much load is required to make the plates slip and therefore allow differential action. The inside tire can be allowed to spin, but that spin is limited by the clutch plates. These differentials have become very popular over the decades. The Detroit Locker attempts the same function, but locks the two wheels together solidly under power. There is no adjustment, and part throttle application tends to be a bit jerky as it alternates between fully locked and fully open. Not too many racers continue to use the Detroit Locker except where it is required by rules as in Winston Cup. The disadvantages of the limited slip, no matter how it is constructed, are that it is rather heavy and must be both accelerated and decelerated. It also requires a fair bit of maintenance to function properly. It is a compromise between the open differential and the spool, and, as such, it's not perfect for all applications.

In the '80s, the torque-biasing differential came into vogue. By using worm gears, this type of differential transfers a larger portion of the engine's torque to the tire that has the greatest traction. These are now made by Quaife in England, and have been produced for racing cars by Gleason-Torsen in the U.S. and may soon be again by Zexel-Gleason. As with all of the other designs, the object of this design is to eliminate wheelspin at the corner exit at the inside tire where load is decreased. Its only real disadvantages are high weight (it's even heavier than most limited slips.) and the friction of all the gears meshing inside the case. The torque-biasing differential does a very good job and would probably take the market from the limited slip, except that other new differential types have recently been introduced. These include all kinds of electrically operated torque-sensing mechanisms, viscous couplings, differentials that use eccentrics rather than gears, and many others. The future will be bright for the company that comes out clearly ahead with the better mousetrap. Meanwhile, when allowed by the rules, I use an open diff for large ovals, a

The open differential is common in all economy cars and sedans and applies torque evenly as long as wheelspin does not occur. When it does, all torque is applied to the tire that has less traction. This is acceptable only when speeds are kept quite high and wheelspin never occurs, such as on long ovals. Many entry-level cars such as Spec Racers, Formula Fords, and 2000s are required to use them.

The Salisbury differential uses clutch plates to couple the wheels, and the load on those plates can be varied by the use of loading ramps of different angles. Different ramps may be required when going from the four-mile track at Elkhart Lake to tighter Sears Point. This means more time working on the car between races, but also means improved performance and lower lap times than when using a non-adjustable differential.

This torque biasing differential made by Quaife uses worm gears to send more torque to the more highly loaded tire. It is adjustable by using gears of different tooth angle, as well as by altering the Bellville washer stack loading the gears. Its only real disadvantage is weight, and the case is now made of aluminum, which helps. It is a very good choice on slower tracks such as street circuits.

spool and perfect stagger for short ones, a limited slip or spool for high-speed road courses, and a torque-biasing diff for low-speed road courses and street circuits.

The type of differential used greatly impacts a car's reactions to the driver's control inputs. Not only can it make the car prone to predictable handling difficulties, for which we must adjust other parts of the car, but control difficulties may arise as well. With an open differential, as required in Formula Ford and Formula 2000 for example, putting an outside rear wheel off the track into the dirt can be a prelude to a spin. A car using any sort of a differential approaching lockup under acceleration is almost guaranteed to spin if an axle, halfshaft, or CV joint breaks. Knowing the characteristics of the type of differential you are using may require adjustment of driving techniques to take advantage of its good tendencies and avoid those that are not so good. It may also reveal reasons for handling problems that your car exhibits and suggest more efficient solutions for eliminating them. As always, other parts of the car must be adjusted to suit the type of differential being used. All of the magic screws must be properly set to work with one another.

Aerodynamic Aids

Aerodynamics is a subject that most racers feel to be rivaled in complexity only by cosmology and elementary particle physics. Air does not always do what we expect, and when it does deviate from our expectations, being invisible, it does not let us know. Fortunately, for those of us who only race cars and do not design them, only two basic concepts are necessary for a reasonably complete understanding of the aerodynamics as it relates to race cars.

Daniel Bernoulli was a Swiss mathematician who lived in the middle of the eighteenth century. He derived an equation which has since been used to design wings for all modern aircraft, ground effect tunnels for race cars, and even the venturi sections of wind tunnels, carburetors, and jet engines. Bernoulli died in 1782, the year before the Montgolfier brothers got the first hot air balloon into the air.

Bernoulli showed a mathematical relationship between the pressure and velocity of a fluid, but he was mainly concerned with the application of these principles to water. It was later realized that all liquids and gases are fluids and obey the same laws of fluid dynamics. Gases behave like liquids with much less density and viscosity. Applied to race cars, Bernoulli's formula says that when air is displaced by a surface, its velocity increases and its pressure decreases.

Some of the Sprint Car crowd favors the concept that the air hitting the top of their wings pushes the car down to the track and increases the load on the tires, thereby increasing traction. For this reason, they run their wings at some fearsome angles, but their understanding of how a wing works is not exactly correct. To racers, at least, the wings on an aircraft are mounted upside down, but in that application, it is easier to visualize the real reason that a wing works. As air flows around the top and bottom surfaces of an aircraft wing, it must flow a longer distance over the top surface than over the bottom. Additionally, the air cannot follow the complete contour of the top surface but separates from the surface at some point forward of the trailing edge. The result is an increase in the velocity of the air on top of the wing, and as Mr. Bernoulli showed us, it forms a bubble of low pressure air above it as well. When a plane is in steady, level flight, it is the low pressure area created on top of the wing that keeps it aloft. The camber of an aircraft wing is increased by lowering the flaps on takeoff, which increases the pressure differential above and below the wing, and is responsible for the increased lift needed for the climb after takeoff. On race cars, where the wings are mounted differently, the lower surface of the wing is pulled downward by the low-pressure area underneath it. The increased download provided by the wings is a combination of a small downward pressure above the wing and a much larger suction below it. Figure 8-4 illustrates the concept.

The first wings on a car, excluding the air brakes of the Mercedes and Auto Unions of the '30s, were used on the Chaparrals of the '60s, and were the idea of Jim Hall. By today's standards, they were pretty crude, being non-cambered (symmetrical), having a slight angle of attack to the ambient air stream, and using no side plates. It was soon discovered, though, that they were already producing more downforce than could be handled by the long limber mounting tubes that attached the wings directly to the rear suspension uprights. Graham Hill (Damon's father) suffered a catastrophic crash in his Lotus at Barcelona in '69 when one of the mounting tubes collapsed. Lotus was noted at that time for only building components as stiff as necessary. Sometimes they underestimated.

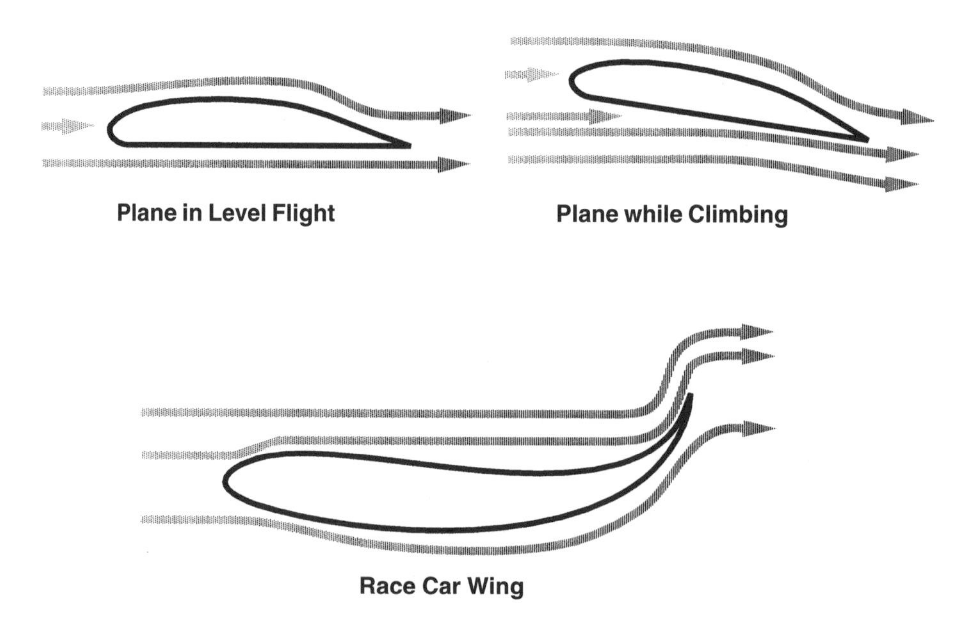

Fig. 8-4 When air is deflected around a convex surface, its velocity increases and its pressure decreases. The suction on the curved surface is responsible for the majority of the force production.

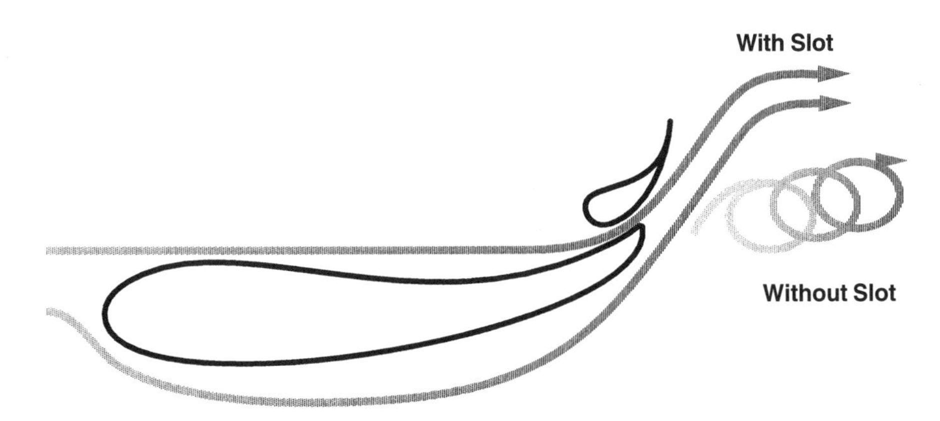

Fig. 8-5 The airflow going through the "slot" created between the elements helps keep the flow from becoming turbulent behind the second wing element at high angles of attack.

It was also discovered that the downforce generated by a wing increases with increasing angle of attack. At some point, though, the wing will stall and produce no downforce at all, but a huge amount of drag. The Sprint Car boys could learn something here. For any given wing profile, angle of attack can be increased to approximately 14 degrees before stalling occurs. The tolerance may be plus or minus two degrees, depending on the profile, but in any case, this is a good rule of thumb to follow. When a wing stalls, the boundary layer can no longer remain attached to the surface, so it separates and the wing produces a huge amount of drag and almost no downforce. In such a case, the only downforce is that generated by the small force pushing down on its top surface.

When racers began to understand what caused the stalling of their wings, i.e., the separation of airflow on the wing's underside at high angles of attack, they slowly realized that a wing of a shorter chord could be used to help keep the flow attached. In order to get the required surface area and angle of attack, it was then necessary to introduce a second element to the wing above and behind the main element. This second element increased the overall area and increased the angle of its chord line. The introduction of new, high-energy air from the top of the first element to the bottom of the second helped to keep the flow attached below the wing, or at least, to reattach it. Since then, this type of wing configuration has become common practice and even three-element wings are proliferating on road courses on cars with enough horsepower to handle them. With multi-element wings, the lift-to-drag ratio has been improved enormously over what it would have been on a single-element wing of the same effective angle of attack. Aircraft use flaps (essentially the same as our second and third elements) for their high lift requirements. Some use Fowler flaps, which are *exactly* the same as our second and third elements. Figure 8-5 shows how this higher-energy airflow helps to keep the flow attached to the lower sides of the additional elements.

In terms of the magnitude of the pressure differential created, the shape of a wing is overrated by some. The door to your house could be mounted on a race car at some angle of attack and would produce downforce, but with an accompanying severe penalty in drag. Since the Wright brothers' first flight, all of the work done on wing sections has been to produce shapes that are more efficient in their lift production with respect to generation of drag. Some very "clean" wings have been produced for aircraft use, where maximum lift is rarely used and Reynolds numbers are comparatively high. For racing use, however, wings are considerably "dirtier," causing more drag than they produce downforce, in many cases. This aerodynamic drag is a physical force holding the car back both while accelerating and at its top speed.

All is not as bad as it seems, though. Since the advent of ground effect technology in the early 80s, many cars have been built which produce some really impressive downforce figures with the attendant abundance of drag. In every case, a car on a road course has been faster with more aerodynamic force gluing it to the ground. This assumes, of course, that the tire companies are aware of the loads being generated. Formula Ford tires on an Atlantic car with wings and tunnels would allow some really spectacular yaw angles, but would hardly be fast. Conversely, on only one occasion have I seen a car that was equipped with too much tire. In this case, the car could be thrown into the turn and would almost come to a stop. The engine made too little power to keep the tires on that proper balance point between traction and spin. *You must make the car dance!* The VDS Can-Am car driven by Goeff Brabham in 1981 and Michael Roe in 1983 and '84 was among the highest downforce-producing cars ever built. It was up to 20 MPH slower down the straights, but held ultimate track records as many as five years after it was last run.

On road courses, it is difficult to have too much downforce, even with the increased drag it inevitably produces. On big ovals, though, the situation changes due to another important concept that every racer should be familiar with: the inverse square law. This law permeates all of physics from optics to—you guessed it—aerodynamics. Both the downforce that a wing produces and its attendant drag are increased by a factor of four when the speed doubles. Put more simply, an Indy Car running 100 MPH and creating 100 pounds of downforce and 100 pounds of drag from its rear wing will produce 400 pounds of each at 200 MPH. When the CART cars run superspeedways, they leave their road course and short oval wings at home. At the extremely high speeds reached on the 2-½-mile oval, the big, high-cambered wings would produce tremendous downforce, but much too much drag. Actually, they would produce so much drag that normal qualifying and racing speeds could not even be reached. Aerodynamic drag plus power lost in the gears, bearings, and rolling friction will equal the maximum power made by the engine at top speed. To increase top speed requires an increase in power, a decrease in frictional losses, or a decrease in aerodynamic drag. Period.

Misinformation seems to abound with respect to the tunnels carried by ground effect cars. The roof of a tunnel is analogous to the lower side of a (race car) wing. It uses a venturi section where the air is forced to accelerate, increasing velocity and reducing pressure. The magnitude of this decreased pressure is enhanced by the fact that the ambient airflow (that airflow normally passing beside the car without being disturbed by it) is discouraged from being sucked into the low-pressure area inside the tunnel by the body sides. In the days of tunnels with skirts that drug on the ground, a very effective seal could be arranged, almost eliminating the dilution of the low-pressure area within the tunnel which sucked it and the car to the ground. One by one each sanctioning body became disenchanted with "ground effects." They first outlawed the skirts that sealed the tunnels so effectively, and then passed rules limiting the amount of curvature the bottom of the car could exhibit in profile. The last of the true ground effect cars with skirts and unlimited profile were SCCA's Super Vee's, which, for a time, made more downforce than any other type of car in the world. Since they made more downforce at high speed than they weighed, they could, theoretically, adhere to an inverted track. If you get the chance to do this, just don't ever slow down! With the passing of the new regulations, though, all classes of ground effect cars became known as flat bottom cars. After several years of development, the flat-bottomed cars are once again making more downforce, at speed, than they weigh; but speed is the key. Since the downforce they generate increases as a square function, a tremendous amount of downforce is lost when running lower speeds.

Since the brass ring always seems to be to find the best compromise for the situation, engineers are constantly playing around with a car's aerodynamics looking for an advantage, however slight it may be. Aerodynamic setup is relatively simple if the car in question is a front-engined car using a front air dam and rear spoiler. It is relatively simple because little can be done with such cars except to take them out to the legal limit. On a car with only front and rear wings it is a bit more difficult because such cars usually have little power and cannot afford much of a top speed penalty in order to have increased downforce and with it increased cornering ability. Some higher-powered sports racing cars such as SCCA's latest Can-Am incarnation (née IMSA's WSC) cars only use rear wings, however, and must balance the downforce they receive from the low nose with that from the rear wing. This is not always easy, not always effective, and sometimes results in a car with handling characteristics that change between low and high speed. The ground effect cars, whether flat bottomed or not, must arrive at the proper balance between downforce and drag, as do other types. The balance between low-speed and high-speed handling can be altered by moving the position of the center of pressure created within the tunnels and by the front and rear wings. The wings can be varied in profile, angle of attack, trailing edge lip, and sideplate design. If we are speaking of cars with tunnels, this means that there are now four additional "magic screws" at each end of the car, plus two more alongside the driver, for a total of ten more "magic screws." (This assumes that you are not running an oval, in which case an asymmetrical setup could increase this number further. Although some look down on ovals, they provide the opportunity to set up a car to a much finer set of conditions.) Fins, tabs, lips, splitters, and spoilers add a few more ingredients to the mix.

Two requirements must be satisfied before a car with aerodynamic aids can be made to be both fast and easy to drive. First, the aerodynamic balance point, the center of pressure, must be located reasonably close to the center of traction. Both the center of pressure and center of traction are moveable by use of various chassis and aerodynamic adjustments. In this way, the car's high-speed balance will be the same as its low-speed balance. After this is done, the balance between downforce and drag must be tailored to the conditions, being careful not to upset the center of pressure location. The objects of this exercise are to make the car corner as fast as possible, accelerate as fast as possible, and arrive at as high a terminal velocity as we can, all while the car remains balanced, predictable, stable, and responsive. Of these, stability and response are paramount. Even if the car is not properly balanced, it should be predictable. The old saying around the paddock is, "To finish first, first you must finish." A corollary of this is: "To be fast, a car must be easy to drive fast."

Alignment

Many racers seem to think that a car should be "aligned" once during a season and that is all the chassis tuning that is required. Nothing could be further from the truth. In reality, the alignment of a car is accomplished through the adjustment of many of the parts we have been discussing, plus a few others not yet mentioned. To be sure, it includes setting the camber and toe, but these as well as every other item included under the heading of alignment are variable and are used to alter the car's handling characteristics as required. An "alignment" should be part of each pre-race preparation, to return to a previously established baseline, or to set up for expected conditions for the next race.

Since setting the toe in or out on a pair of wheels is what most people think of when they hear the word "alignment," this is a good place to begin the analysis of the components of a real alignment. These components also include setting camber, caster, bump steer, ride height, and corner weights. When the steps that make up an alignment are actually performed on the car, the setting of some will change values of others. This will require that the process be done in a certain order, depending on the configuration of the car in question. But some amount of tail-chasing is inevitable, regardless of the order in which the alignment steps are performed.

In the early days of racing, toe in was preferred at the front of all cars. Since the forces acting on the tires when traveling straight ahead are outboard of the wheels' pivot points, the ball joints, they naturally work in a direction to produce toe out. Thus, compliance of the steering and suspension bearings in those early days required a bit of toe in so that the outward forces would, theoretically, bring the toe setting to zero. Current racing cars use spherical bearings for the suspension and steering, which are much more precise. Additionally, the stiffness of the suspension and steering parts is now much greater than in the early days of racing, and we no longer must use toe in. Actually, to combat the ever-present corner entry understeer, we have found *a little* toe out to be quite effective. But note the emphasis on the phrase "a little." Too much toe out will cause the car to wander on the straights, and can cause violent darting while braking. Toe is not measured in sixteenths of an inch, but in thousandths of an inch or in minutes (one-sixtieth of a degree) of angle. Toe out should never be used on the rear. A little toe in there adds stability.

When a car is set up using Ackerman steering, the mechanical linkage of tie rods and steering arms produces a condition in which the inside wheel turns farther than the outside with steering wheel movement. Other methods can be used to produce the Ackerman effect, too. In theory, its purpose is the same as that of the differential. Since the inside wheel follows a shorter radius than the outside in a turn, it should turn through a sharper angle. This amounts to variable toe out. In reality, the front tires are operating at much different slip angles due to the weight transfer between them, which seems to negate the requirement of the inside tire turning farther.

For a long time, many engineers resisted the notion that Ackerman steering could be beneficial on any kind of racing car. That thinking changed, though, when it became apparent that Ackerman steering could reduce the ubiquitous corner entry understeer. It seems that what is happening is that in the instant after the steering wheel is turned, but before significant weight transfer takes place, some benefit can be derived from the inside tire turning at a sharper angle before all of its normal load goes away. Engineers now feel that this extremely short time is critical in terms of corner entry understeer. As every experienced driver knows, it takes a finite and seemingly much too lengthy period of time for a car to recover once understeer has begun. It is likely that this first instant of turn initiation is critical to the entry phase of the turn, and some Ackerman effect at this time can be beneficial. It is of no use whatsoever after significant weight transfer has occurred.

How much Ackerman should be used? As always in chassis set up, the answer is ambiguous. Not much is usually required. This assumes, however, that no other serious problems show up to cause significant amounts of corner entry understeer. If they do, increased Ackerman can sometimes be used as a band-aid. I use zero Ackerman on a new car, but carry additional tie rods and steering arms for those occasions when they are needed.

Bump steer is the geometric reaction of the steering to suspension movements. On a double A-arm independent suspension, a wheel will move inboard somewhat as it moves either up or down. That ratio of movement is not equal with direction, and can also have a variable rate depending on the vertical position on the wheel at which it is measured. Since the tie rod must be mounted at some finite height, some relative lateral motion between the suspension and steering is inevitable. When a bump steer adjustment is made on a car, it is (usually) to reduce the amount of toe change as a function of wheel movement. Sometimes a toe change with that movement, called bump steer, may be just what the engineer ordered for a given condition. Knowing when this may be beneficial comes with experience and trying things to see how they work. Some toe out in rebound may be a good thing at the front to reduce corner entry understeer. If it is used along with static toe out and some Ackerman, the tire drag on turn entry can be tremendous. Toe in during bump is sometimes used on the rear to promote stability through the turns.

As mentioned already in the discussion of tires, camber is usually set by examining tire temperatures. The goal is to attain the same temperature at the inner edge, center, and outer edge of each tire. Tire temps can be deceiving, though. Tread rubber cools quickly, especially on open wheel cars. The temperatures seen on a pyrometer in the pit lane are but a vague remnant of the heat that was generated in the last turn the car negotiated. For this reason, a skid pad is the best place to check tire temps, with short ovals coming in a close second. Taking tire temps at a track such as Road Atlanta is almost useless. There, the last turn before pit lane that the driver can take at speed is Turn Seven at the entrance to the straight, which is almost a mile from the pit area. Tire temps can still provide at least some information in these situations, though, even if it is only about tire pressure or compound.

A tire temp that is too hot at the inside edge means too much negative camber. The opposite obviously means too little. A center temperature that is too hot can mean either too much pressure, which crowns the tire, or too little, which allows the centrifugal force to sling the center of the tread out. Some specific camber angle may be set in the shop, but it should not be regarded as absolute. Change it at the track based on the tire temps.

Every car has a designed ride height at which it theoretically works best. Real-world conditions prevail, though, and a car should really be run as low as is practical at a track without bottoming hard. This means some adjustments of ride height at an event. As ride height is changed on an independent suspension, the camber also changes and the toe may change with camber. This means continued compensatory changes to the camber and toe settings as well.

Constant camber changes are sometimes required for other reasons, too. I once went to a race at the Phoenix oval with a driver new to ovals. Checking tire temps, I set what appeared to be the proper camber settings. Rechecking tire temps after the second session showed the camber still insufficient on the outside wheels. After again changing the camber, the third session still showed it to be insufficient. Finally, I realized that the driver was getting used to the track and driving harder, causing the chassis to roll to a higher angle in the turns. This made additional camber necessary.

Caster is the angle by which the lower ball joint leads the upper one on the front suspension. Too much emphasis is sometimes placed on caster. For road racing cars, the two sides should be within 1/2 degree of each other. High amounts of caster (eight degrees or so on most cars) will reduce the positive camber gain on the outboard wheel, which is built into the suspension by the steering axis inclination angle (king pin angle) when the steering wheel is turned. The disadvantage of using this much caster is that the steering becomes quite heavy, especially on a heavy car without power steering. On ovals, some caster stagger is used, with the right side given more caster to allow the car to turn left more easily.

Corner weights are neglected by most club racers, but Mark Donohue considered setting them to be the most important part of proper alignment. Their importance is evident when the tires' dynamic responses to the loads put on them are considered. The best method of setting corner weights is through the use of digital wheel scales and a flat pad. These are in common use by the teams that can afford them. Platform scales (grain scales) are cheaper and adequate, but bathroom scales are not. A mid-engined, center seat car should be set with no more than five pounds difference between the left and right corners. Offset driver cars will be more difficult to bring into this tolerance, but your efforts to do so will be rewarded with a car that is more easily balanced and easier to drive, and that produces lower lap times.

Racers have used many different methods to perform these alignment adjustments, which vary with the configuration of the cars and the budget, experience, and inclination of those doing the work. Use whatever method works for you. Indy cars at Indy are aligned using computerized equipment made specifically for the job by the tool manufacturers. The rest of the time their crews depend on alignment strings. These are ordinary strings pulled tight down each side of the car parallel to its centerline and to each other. They are used as reference planes for setting toe and track width. Camber and caster are set using digital inclinometers. Special alignment wheels are sometimes used for this purpose, and can be very skinny wheels with narrow tires mounted on them, but more often are just aluminum discs replacing the wheels and facilitating measurements. All of these adjustments are made on a leveling platform, a piece of equipment that has become essential for ride height and corner weight measurements at the track and in the shop.

Using Data Acquisition

A technological revolution occurred in the '60s. This was a time when racers were becoming aware of the existence of some of the tuning devices at their disposal and began to understand the relationships among them. The Dutch company Koni was the first to show us that a car's handling could be improved through adjustable shocks. At that time, the constructors such as John Cooper, Jack Brabham, and Colin Chapman were beginning to understand the relationships among springs, anti roll bars, and weight transfer. Tires grew wider, and at least a little stickier, and more sensitive to changes in chassis adjustment. The more we learned, though, the more complicated chassis setup became.

The basic elements of chassis setup, springs, bars, camber, and the like, were relatively easy to use to balance a car with a seat-of-the-pants approach. When finer adjustments became important, though, it became necessary to determine what the cars were really doing rather than the well-intentioned opinions of the drivers as to what they *thought* was going on. Since a modern race car has literally dozens of "magic screws," each of which has an effect on the function of the car, and since the adjustment of each of them depends on the adjustment of others, it can be very difficult to determine exactly why a car behaves as it does and what should be done to improve it.

In the '80s, another technological revolution produced a quantum leap in understanding and with it, performance. This was the widespread use of on-board computers or data acquisition systems. A data acquisition system is, up to a point, the perfect driver. Although it cannot (thankfully) control the car, it can and does remember perfectly every nuance of the car's behavior. With a data acquisition system, the driver must no longer remember his RPM at the entrance and exit of each corner, he is not allowed to forget the oversteer that occurred through Turn Four, and he cannot deceive the engineer, intentionally or otherwise, about when he opened the throttle at the exit of a particular turn.

A data acquisition system consists of a small computer that records information from sensors placed at strategic locations on the chassis. A typical system uses a two-axis accelerometer which will plot a map, using the accompanying computer software, of exactly where the car was at any point of a lap. Additional sensors are attached to the shocks, which indicate shock travel at any of those points. With the help of the software, the normal load on any tire at any time is calculated. Additional sensors show engine speed, wheel speed, steering angle, throttle position, ride height, tire temps, and as many other parameters as can be imagined. When this data is downloaded to a laptop computer, it can be analyzed by plotting various parameters against each other, making a variety of balance, stability, response, and cornering power problems evident. As always, though, it is the experience and insight of the person doing the analysis that reveals such problems and the solutions to them.

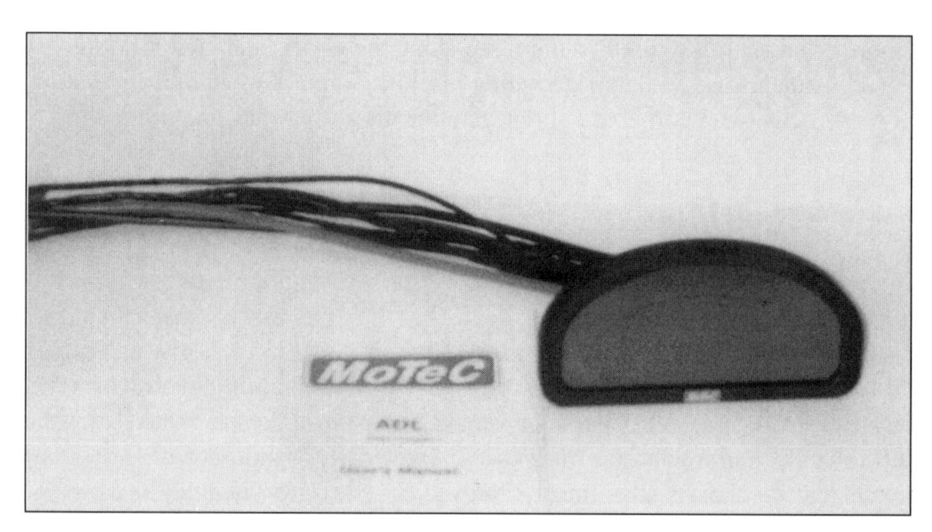

Data acquisition computers are no longer the black boxes they were only a few years ago. They now double as the instrument panel and have gotten noticeably smaller and lighter like this one from Motec. The only other components required are the sensors and wiring harness. With a system such as this, areas of performance improvement can be identified so that the proper chassis adjustments can be made.

A racing engineer must play detective in solving handling problems—by looking for clues at the scene of the crime. Hidden within the data are clues that will reveal why the chassis acts in the manner it does. However, the data generated from even the simplest of recorders can overwhelm the driver or engineer new to this technology. As in other facets of racing, a systematic, step-by-step approach is the key to unlocking the door to understanding.

As an example, let's look for the cause of the oversteer a driver might have reported in the entry phase of Turn Four at a typical track. Looking at the plot of lateral acceleration, it is easy to spot a problem area, indicated by a significant drop in lateral Gs in that part of the turn. This drop does not by itself indicate oversteer, but this is the problem the driver reported, so it is the suspect. By overlaying a graph of steering angle for the same portion of the track, we see in Figure 8-6 that the driver countersteered at exactly the same time as the drop in lateral G loading. These two indicators taken together almost always indicate an oversteer condition. When two or more graphs show concurrent anomalies, the patterns are said to be signatures of particular problems. Quick and proper diagnosis of chassis problems by computer, then, requires that the engineer be familiar with the common signatures.

We have now found exactly where the problem occurred, but not its cause. To do that requires looking for additional patterns in other graphs which occur at the identical spot. Since the problem arises during the entry phase of the corner before the throttle has been opened, wheelspin is not an option. Too much rear brake bias could be the culprit, but the oversteer condition happens so late in the entry part of this corner that it is unlikely to be the cause. With these two common causes ruled out, the engineer might decide to look at a plot of the outside rear shock movement. In this case, a shock plot might show severe and rapid vertical movements of the wheel, indicating a bumpy section of the track on which the shock was not properly keeping the tire nailed down. What this would indicate is that, with the tire momentarily unloaded, the rear of the car stepped out, requiring the driver to make a steering correction. With this knowledge, the engineer could verify with the driver that the entry to Turn Four is indeed bumpy and make the necessary changes to the outer rear shock stiffness to solve the problem.

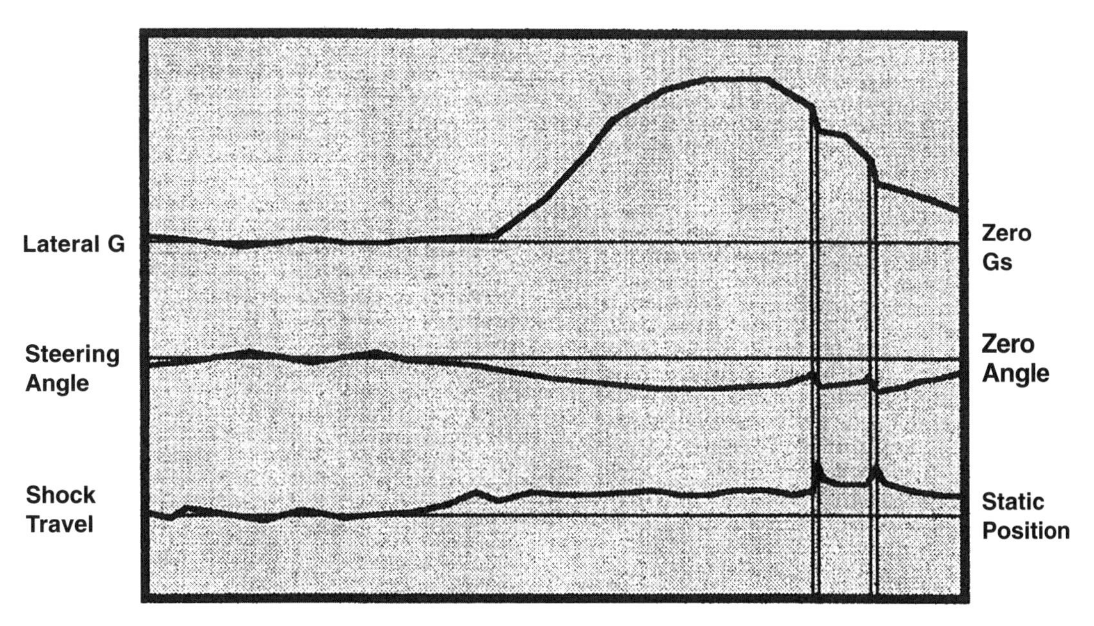

Fig. 8-6 An examination of these graphs, as discussed in the text, shows steering corrections at the same times as drops in lateral Gs, which are signatures of oversteer. By overlaying a plot of shock travel of the outside rear wheel for the same turn, it can be seen that the track is bumpy here and the shock is not doing an adequate job of controlling wheel travel.

When a particular problem is not evident, the engineer's task is to maximize speed at any portion of the track. The easiest way to do this is to study plots of lateral and longitudinal Gs and speed. Anytime a dip is encountered in an otherwise smooth buildup of G forces in cornering, braking, or acceleration, a problem is indicated. By overlaying graphs of other specific chassis functions, clues may be found which indicate a problem that no one even knew about, but that is keeping the car's potential from being realized. The speed graph and lap times are the final report cards. Anything that increases the car's speed at a given point on the track without hurting it enough at another point to cause an increase in lap time is beneficial. After identifying a specific problem on the computer and making an adjustment to the car to improve its function, always double check the effectiveness of the change by studying the new speed graph.

In addition to searching for clues to "handling crimes" committed by the chassis, a good racing detective (the engineer) must also keep his eyes open for any other oddities or patterns that look interesting. This opportunistic approach means not looking for anything in particular, but keeping your eyes open, such as when you find a cracked rotor when changing brake pads.

A multitude of problem areas may be identified in this way, from incorrect tire compounds and wing angles to improper driving technique. Additional information may be gained on engine performance and the most efficient shift points to use. The ability to use precise, detailed, and accurate information to set up today's race cars has become essential, as the cars have become more complicated and more sensitive to adjustments. Just as no businessperson can now afford to be computer illiterate, every racing engineer and driver must be comfortable using data acquisition.

Proper chassis setup is essential to making a car fast. Provided the engine makes adequate power, the gear ratios of the transmission are close to optimum, and the tires are of the correct compound and not old or hard, the most critical aspect of making a car fast is the adjustment of the springs, bars, shocks, alignment, wings, etc. The engineer must configure the car to complement the design of the car, the driver's style, the nature of the track, and the weather conditions. That setup may also be altered to achieve a faster qualifying lap, to accommodate a specific race length, or to enhance a particular race strategy. So many adjustments are involved in the setup that most amateur racers are terminally intimidated and fail to ever learn any more than just the basics. This leads to an advantage for the racer who will take advantage of others' ignorance of the subject by learning vehicle dynamics himself. To move on into the league of professional racing requires that the driver himself, as well as some of his team members, be proficient in this area.

Chapter 9

Getting Sponsors

When we receive financial considerations from a sponsor, we should not be in competition with the other advertising media, but should sell our medium, racing, as the cornerstone of a cooperative ad campaign using the traditional media.

How Sponsorship Works

It is amazing how few amateur and semi-pro racers have a proper view of what sponsorship is all about. Some teams go to a lot of expense and effort to land a sponsor, only to spend their money foolishly or neglect the sponsor's interests and not be able to keep him or attract another. Other racers go to the opposite extreme and seem not to be comfortable with asking a sponsor for money, thinking that they have nothing to offer, or that they are only asking for a handout. Still others fall somewhere are in the middle and ask a potential sponsor for the wrong things or do so incorrectly and fail to land him. With any of these groups, it is clear that a shift in thought pattern is necessary. Sponsors can be attracted at any level of motorsports, even club racing. Having spectators at your particular kind of racing is not a prerequisite for obtaining a sponsor. Knowing what a prospective sponsor wants, though, can give you a clue as to what you should do attract him.

Unless a sponsor is an Angel (more about them later), he is interested in increasing his bottom line. Whatever his product or service, he wants to sell more of it as a result of his involvement with you in racing. Part of increasing sales is the sponsor's responsibility, but some of it belongs to the racer, too. When we receive financial considerations from a sponsor, we should not be in competition with the other advertising media, but should sell our medium, racing, as the cornerstone of a cooperative ad campaign using the traditional media. If the advertising the sponsor buys from us is not effective, he will not be back. Even worse, his bad experience may result in other teams who operate in a more professional manner not bringing him back to the table.

If the sponsor's business is a very small one, such as Joe's Garage or Bob's Tire Store, the exposure he receives will probably be in the form of his name on the car and a photo of it on his office wall. Periodically during the season, he may want to display the car in front of his place of business to help attract customers. Payment for this type of advertising is usually in the form of a discount on parts or perhaps shop space for you to work on your car. If your racing expenses are low, such sponsors can be a great help in keeping your budget in the black. If you can attract a sponsor who will paint the car in exchange for putting his name on it, another who will give you a discount on racing tires, one who will allow you to use a trailer he builds, and another who gives you oil and racing fuel during the year, your budget can be cut considerably. Travel expenses can be bartered in this way, too. Hotels or motels sometimes give free rooms to racers who advertise for them.

Sponsorship works at any level. The sponsors of this amateur car include distributors of oil and related products, an engine builder, racing parts suppliers, a real estate company, and driver Mark Weber's own business, Exclusive Sports Car Photography.

With very small businesses like these, accounting practices for determining the increase in sales due to these types of sponsorships are usually nonexistent. This means that you must be very sensitive to the sponsor's mood swings to keep him happy. Go out of your way to try to bring him business. If you agree to put your car on display at his place of business for six hours on his busiest day, do it for eight and be there with it talking to his customers and handing out his brochures. Make the sponsor think he is getting more than his dollar's worth. In the absence of any type of cost-to-benefit analysis of sponsorship, the sponsor's perception of the value he gets for the service you provide is paramount.

Larger business sponsors will net you a larger check, but you should be prepared to use a different approach. For businesses that produce expensive products or those normally distributed in large quantities to only a few distributors, the approach you should emphasize is corporate entertainment. This is also called courting the buyer. A manufacturer of, for example, plastic injection molding equipment may sell to only 20 or 30 molders total throughout the country and his products may range from $80,000 to $250,000. If you entice him to support your racing program, you will need to help him entertain his clients by taking them to see you race. Usually some sort of a hospitality tent is set up in the paddock, which houses a caterer and tables for meals and drinks for the sponsor and his clients. (Good, hot meals are also welcomed by the crew. Be sure the drinks are only consumed by the sponsor's people, though. Some tracks and sanctioning bodies strictly prohibit alcoholic beverages in the paddock.) The clients are introduced to the driver and are photographed with him. They are also introduced to any celebrities who may be present at the event. In short, the sponsor's clients are treated like racing insiders and given a view of racing that they could not have seen on TV or from the stands. It is this feeling of being an insider, of being someone special, that the sponsor wants to leave with the client. The idea is, the client will remember the event and who made it possible, the sponsor, the manufacturer who builds the product he wants to buy.

Now let's look at a slightly different example. Let's say that you are the manager of a restaurant chain and are approached in your office by a beer supplier. He asks, "How are you, Mr. Smith? How are the wife and kids? How is the golf game? How many cases do you want this month?" Then you are approached by a salesman from a competing company. He says, "How are you, Mr. Smith? How are the wife and kids? We are sponsoring Adrian at Indy this weekend. We would like you and your family to be our guests at the race." You go to the race with your family. Your kids get autographs from some of the drivers. Your wife meets Paul Newman, and you all are treated like royalty. The next week the same salesman comes in and asks, "How are you, Mr. Smith? Did you enjoy the weekend? How many cases do you want this month?" Which salesman will get your business?

Corporate entertainment is also frequently used by manufacturers of consumer products to entertain their distributors, but can be combined with an advertising campaign aimed at the consumer. This approach uses media and print ads, using the sponsor's involvement in racing as a cornerstone. Several forms of advertising media are now used in a synergistic manner to generate better results than any one of them alone. A long-distance telephone company, for instance, may emphasize the speed with which calls can be placed by using photos of your car in its print ads. In selected areas, the company may advertise on radio and TV that its car will be competing in a particular race. Another example might be a soft drink company that wants to suggest that its product is bold and exciting. In this way, the company's involvement in racing is integral to their advertising.

For any sponsor, a self-leveraging plan should be organized. This means that, worst case, the first year of sponsoring your team should cost the sponsor no money, and every year thereafter they should make money. Integrate your racing into their advertising campaign to increase sales. This may be done in a variety of ways, but the most common these days is by combining sponsors, i.e., by bringing together your different sponsors in business deals. Suppose you are talking to the shipping giant, UPS, about sponsoring your Winston Cup team. What products and services do they use? They buy trucks, planes, tires, petroleum products, materials handling equipment, uniforms, packing materials, computers, and office supplies. Now suppose you talk to GMC about sponsoring your team and arrange for them to sell trucks to UPS. In exchange, UPS gets GMC's parts shipping business. Additionally, you convince IBM to sponsor your team by arranging for them to sell computers to UPS. Who gets the contract to ship all of IBM's products? UPS, of course. You get the idea. A decade ago, a team paid its operating expenses with its sponsorship dollars and any money won was profit. Today, by combining sponsors, a team can make a large amount of money by landing more cooperative sponsors. This method was pioneered by the Formula One constructors, but has since found its way into most top-level motorsports worldwide. Using this method, a smaller team can generate the resources necessary to grow into a large, well-funded team.

Regardless of the type or size of business you approach, the very best method of landing the company as a sponsor is to find an ally on the inside. In *Stalking the Motorsports Sponsor,* Pat Bentley tells us we should find a "closet racer." This person is high enough in the company to exert some influence and is a racing enthusiast. If you can convince him that racing is a viable ad medium, and that you can carry his company's banner, he will be your champion and will convince management that their bottom line can be increased by its involvement with you. This amounts to making a single person perceive the value of advertising through motorsports, just as you would in trying to secure the sponsorship of a small business such Bob's Tire Store. You then use that person to lead your campaign inside his business. To find such a closet racer, you might try a method that was successful a number of years ago. In this method, an ad was placed in a business magazine featuring an Indy car with a caption that read, "Can this car make money for you? It did for Citicorp." The ad generated over 20 responses, five of which were deemed serious. Of these, one generated a signed contract.

The businesses you seek as sponsors will always have allotted advertising budgets, which is all they will spend for a fiscal year. When you ask them for money, then, you should be prepared to show them, in detail, how their bottom line will be better by giving some of their budget to you rather than all of it to the traditional media. To do this, you will need a sponsor proposal.

Sponsor Proposals

Whether a proposal is directed to a potential sponsor or to the woman you love, the purpose is the same. It should convince the other party that you are the one for them. Few sponsor proposals are excellent. Most range from amateurish to ridiculous. Some of these have been prepared by professionals in marketing and delivered by professional racers. But experience in either of these fields does not necessarily qualify one to prepare a good proposal.

Most of the bad ones share the same qualities. They talk about how great the driver and team are and what successes they have had, and totally disregard what the sponsor will get for his money. A properly done proposal is a sales tool. To be persuasive, it must show:

1. A need for change from the present advertising method

2. The benefit the sponsor will receive for making that change

3. The recommended solution (advertising through racing with your team)

4. Supporting details to show how the solution will be implemented

Another quality of a good proposal is focus. Ideally, the proposal should target the needs of a specific business. But unless your team employs a full-time sponsor hunter, the time required to do this, i.e., to research the needs of a company and write an initial proposal specific to that business, makes this impractical. Instead, consider tailoring a proposal to a business type. DO NOT write a generic proposal to be sent out to everyone. If you do, your proposal will look like a solution in search of a need. The key to landing a sponsor is to get away from your own interests and focus on the sponsor's needs. To do that, you will need to know something about the business or the type of business that is your target.

If your prospective sponsor is a small- to medium-sized business, they have grown to be the size they are because they are smart and have good business sense. You will not be able to baffle them with BS, so don't try. In preparing your proposal, it is essential that you do your homework. If you know someone who works in the industry, ask what type of consumers make up the prospective sponsor's client lists. Ask what size the average sale is. Find out the size of their advertising budget and how it is currently being spent. If you do not know someone who works in that industry, find someone in advertising who can help with this information. Be careful, though. Some advertising people, like many of the rest of us, are resistant to change and may not feel that racing is a viable ad medium. If you do not know something about the prospective sponsor's business when you prepare the proposal, he will think you are out of touch and do not understand his problems. To make him receptive to your proposal, you must instill confidence in him that you know what you are doing. Remember that the first thing your proposal must do is to convince the prospective sponsor that his present advertising method contains a problem that you can solve.

The proposal should take a two-pronged approach. The first part should be short, no longer than five pages, to get and keep the reader's attention. This is the bait on the hook with which you try to interest him. The reader is likely to be a gatekeeper whose job is to find any reason whatsoever not to recommend your proposal to the decision makers. To get it past him, show him the benefit that his company will receive by adopting your proposal, make it clear and concise, and back up your recommended solution with facts to show that it will work.

One proposal that meets these requirements came from a dirt track race team. It was geared to western wear stores and is shown in its entirety, as an example, on pages 203–206. The form of the initial proposal should look something like this one. Keep it short, but address all four of the important concerns: a need for change, the benefit of making that change, a recommended solution, and how that solution will be implemented. Use italics, underlining, boldface type, color or any other attention-getting device to emphasize the important points. There is no need, at this point, to include information about your successes, the people making up your team, or demographics. Print it using a laser printer and use high quality paper. Spiral bind it or bind it in a suitable cover, as you would any other business proposal, marketing plan, or business plan.

The proposal should contain a financial page that tells the prospective sponsor the range of financial commitment you are asking for. When preparing your financial page, don't ask the sponsor to make you rich. Your job is to help make him rich. His is to let you race. Selling your needs short should be avoided, too. After the contract is signed, it is too late to go back and ask for more. Be realistic about the amount of money you ask for. When a relatively small amount of money is involved, it may be acceptable to ask for individual sums based on the entertainment facilities available and the position of the sponsor's name on the car, as shown in the financial page on pages 205–206. However, avoid having your proposal look like a prepackaged arrangement. You should sell your program to the sponsor as a customized effort that will suit his needs *exactly.*

People perceive ideas in a variety of ways based on their previous experiences and their personality types. As you write, think of how different people may interpret your text. Avoid any negative connotations of your ideas. A persuasive proposal should be carefully worded to avoid anything negative associated with your plan. As an example of wording that could carry a negative feeling, consider this: "In times of financial recession such as these, when people's discretionary income is low, the marketplace becomes even more competitive. In order to attract consumers, businesses must trim the fat from their advertising to make it *effective.*" That first sentence makes the prospective sponsor worry that he should be very careful about doing any advertising at all, let alone anything different, which is exactly what the racer does *not* want to say. That sentence should have been deleted, leaving only the admonition that the fat should be trimmed from the advertising to make it effective, which is what the racer wanted to get across in the first place. If your program contains an inherent obstacle, address it squarely, up front, and show how its effect can be diminished. Don't try to sweep it under the rug because this will raise a red flag with the reader. If this kind of writing is not your strong point, take your rough draft to an English major or teacher, or an advertising copywriter for a final cleanup.

Some sanctioning bodies and race organizers have videotapes available showing their cars in the paddock and on the track. These tapes can be customized with your team name and logo, and make very good attention-getters when sent with your initial proposal. Use a tape only as an attention-getter, though. A picture may be worth a thousand words, but your proposal is the sales tool. If a tape is not available for your type of racing, professional studios can produce one for you, but the cost is likely to be several thousand dollars. With the advent of good-quality videotape cameras and do-it-yourself editing shops, a professional-looking video can be produced at minimal cost. To get enough quality footage of your car, however, be prepared to spend most of the season with a camera (and someone to operate it) in your pit. Regardless of which video production method you choose, be sure the finished product looks professional. If it does not, it will lead the potential sponsor to think that you will not present a professional image of his business, and that will do more harm than good.

If your proposal gets past the gatekeeper, it will be forwarded to one or more decision makers. If he or they show any interest in it whatsoever, you should meet with them to discuss the second part. At this meeting your job is to convince them that what you propose will work, that it will increase their bottom line, and that your team is the only one that can accomplish it. If you succeed, they will ask you to write the second part of the proposal. This is where you get to research the company in detail to evaluate how you can best design a program to meet their needs. The company's employees will assist you with facts and figures and with exploring avenues that you can use in carrying out the plan. This is sometimes a long and exhaustive process, but time spent on it will yield large dividends. Obviously, the more money you are asking for, the more time should be spent on this research.

Unless you have a background in advertising or marketing, it is best to recruit help in this phase of the sponsor hunt. Several marketing agencies around the country perform this service for racers exclusively, but many others cater to athletes and sports figures and do essentially the same thing. Even a representative of a local advertising agency or marketing firm may be able to help. If a large sum of money is at stake, choose carefully. Your marketing representative is helping to sell you and your racing program and should be personable, have good social skills, and be a good salesman. The same goes for you.

While working within your potential sponsor's company on the best way to increase his bottom line, you will encounter people of four different personality types. These four personality types can be characterized as detail-oriented, pragmatic, consensus-oriented, and visionary. In trying to generate positive attitudes about your project within the company, individuals of each personality type will require different approaches.

Each of the people you encounter will also have different levels of expertise in both advertising and auto racing, ranging from expert to completely uninformed. You should determine each person's level of knowledge in both of these areas. Do this for all of the people you must work with. It is unlikely that any of them will have more than a modicum of understanding of auto racing. But, if you find someone in the company who is an avid racing fan, he is the "closet racer" you have been looking for. It is important to identify the personality type of each person so you will know how to deal with them. You will also meet decision-makers whom the others report to. It is essential to use your best salesmanship on everyone. Make them like you and *want* to recommend your proposal.

The second part of your proposal contains the bulk of the details showing how the goals will be met. The company will set specific goals for their involvement with you in racing. You must work with their advertising and marketing people to ensure that those goals are realistic. You will also work together on specific plans to reach those goals. Checkpoints will be established along the way to make sure the plan is on track and to evaluate its effectiveness. If you do not have a background in these matters, don't panic. Your marketing representative should be well-versed in these areas, and the company's people know what they are doing as well. When all the research is done, however, it will be up to you and your marketing people to write a proposal that details a plan, including specifics, to meet these goals. Remember that the purpose of this proposal is to sell your program. The contract is not yet signed.

The form of this proposal will be a little longer and a bit more complex than the initial proposal. Depending on the amount of money involved, it should be between 10 and 50 pages. It should contain a table of contents followed by a summary of the plan. That summary should be followed by a detailed analysis of the goals that have been set and how they will be achieved. This will include sections dealing with how the point of purchase and media advertising will be dovetailed with your team; how, when, and at which races or promotions clients will be entertained; demographics and media exposure of the sponsor's team through TV and print media; and profiles of the team, the people who will carry out this plan. Do not neglect this last item. Business people know that regardless of how strong the plan is, it is the people behind it who will carry it out. Show them that you are serious, professional, hard working, and straight up.

The proposal should also include financial pages which show not only the amount of money spent with you on racing, but the funds spent with other advertising media as well, and cost-to-benefit analysis procedures to ensure that the company's goals are being met.

Angels

In amateur and entry-level pro racing, sponsorship usually comes from someone you know. This can be your father, your employer, a friend with a thriving business, or someone you can convince to help you from the goodness of his heart. Angels do not expect to see an increase in their bottom line, but they do expect to see results. They help because they want to, but they also expect to see your finishing positions improve as a result of their help. If that does not happen, your Angel will become disappointed in you and wonder why he should continue to help.

An Angel will probably not be interested in your proposal. With him, a more personal approach is needed. If he is interested in your racing program, he will hang around your car, ask questions, and perhaps even show up at a race. This is a very good reason to always look professional. Whenever the Angel is around, you should use your best Dale Carnegie personality (How to Win Friends and Influence Your Sponsors). An Angel usually begins his involvement slowly by buying a set of tires, for example. Next, he may pay for an engine rebuild or a new trailer. Before you know it, he may be supporting your move to a faster and more expensive class and one that will better promote your career. He may also be able to help with business expertise or be able to help you land a corporate sponsor. Sponsors such as these have sometimes gone on to be heavy-hitters in racing, for instance, Jim Gilmore, Paul Newman, and Ron Hemelgarn. An Angel will help you because he likes you. Always remember that he does not *have* to help. Don't miss an opportunity to show your appreciation.

Corporate Sponsors

Corporate sponsors are much like large versions of small- to medium-sized business sponsors. The same procedures and methods should be used to obtain a corporate sponsor, but you should be aware that professionalism is of prime importance, and things will take much longer than you ever thought possible. Big business is solely concerned with how much their involvement with you will increase their bottom line. They will analyze your program to death to find out exactly what the cost-to-benefit ratio is, and they will do it before they decide to do anything else.

To land a big-time sponsor (and here we're speaking of businesses the size of AT&T or Proctor & Gamble), you will probably need to hire a professional sponsor hunter. Such individuals deal with product endorsements that net major athletes millions, and, as such, are in a prime position to handle these endorsements for you. This puts you in direct competition with Troy Aikman and Michael Jordan. It also puts you in the ring with Roger Penske and Richard Childress. If you have gone far enough in racing that you need millions, you will be on the track with these teams, too.

Your sponsor hunter will prepare your proposals from his office and meet with potential sponsors himself. For each business he contacts, he will prepare a proposal just for that business, and he will be familiar with all aspects of the business. At some point, your presence will also be required in the boardroom, and, if the sponsor is very interested, the Board may want to inspect your operation, perhaps at your shop as well as at the track. An agent to handle these details will not come cheap. Professional sponsor hunters usually require a retainer of tens of thousands of dollars up front as well as 10 to 15% of the gross, which the sponsor contributes. If this makes you think they are well-paid, you are right, but you must also remember that it takes months of preparation and meetings and extremely good sales skills in a big-business environment to make these deals happen. Your agent will be well-paid, but he works for a living and the benefit to you will be well worth your investment. It will allow you to race as you want to. Some operations, such as Penske, have a professional sponsor hunter on their payroll whose job is to keep existing sponsors happy and keep new sponsors in line to be employed when they are needed. When you get to this level, this is most likely a cheaper way to approach the problem.

A very small sponsor, such as one you might approach for a set of tires for each race, may make a decision on the spot. But as your needs for racing capital rise, anticipate that sponsors will take longer to reach their decisions, and be prepared for a greater number of rejections. The reasons for these rejections may be obvious, or you may never know why you were turned down. If someone gives you a reason, consider it logically and use it to improve your approach. Sometimes the reason has nothing to do with you or your team. I was recently involved in approaching a sponsor who is currently involved in motorsports with a number of other teams. We were turned down because those teams are not currently winning. The sponsor wanted to get results from those teams before they added a new one. When you receive a rejection like this, it only means that the timing was not right. (Remember—"Timing is everything.") Keep abreast of developments and try again later.

Taking Care of a Sponsor

As you are trying to attract a sponsor, keep in mind what he needs from his association with you and you will have a good handle on keeping him after your initial contract has been fulfilled. His reason to be involved in motorsports at all and, in particular, with you, is to make money. He does that by boosting the egos of his clients so that they will buy his products. He also uses the entertainment of his sales staff to improve morale so that they will be better prepared to make sales to his clients. Additional benefits that he may see are in using racing through his other ad media to increase consumer sales. Anything you can do to increase results in these areas will be beneficial, both to him and to you.

If corporate entertainment is your sponsor's major thrust, the normal hospitality tents and caterers will be the major focus at the track. Remember that he wants his clients to feel special. If you bring someone to the track whose sole job is to roll out the red carpet at each event, it will go a long way toward satisfying the sponsor's needs. This person could be your wife or girlfriend or the advertising person who helped you land the sponsor. Whoever handles this task, be sure that the sponsor and his clients are made comfortable and happy. Your interests as a racer take a back seat to this. Your sponsor's interests come first; whether you win the race or not comes second. Of course, winning the race or doing well makes the sponsor happy, too, so don't neglect your on-track performance. Just remember that your sponsor funds your project. If he is not happy, you do not race.

1994 was a dismal year for Bobby Rahal. The new Honda Indy engine was encountering teething problems and the ill-fated Truesports chassis he was using compounded the problem. He even failed to qualify for Indy that year. The next year found him in a Lola chassis with Mercedes power. In explaining the switch and the apparent lack of loyalty to Honda, Bobby said this, "When you're dealing with a company like Miller Brewing or any major sponsor that's expecting results both on and off the track, you can only expect them to be patient so long." Rahal-Hogan Racing was receiving significant help from Honda, but their primary sponsor was (and still is) Miller Brewing. Rahal did what was necessary to get the results his primary sponsor expected.

Besides having an operation that is competitive and is turning in good results to your sponsor, your image at the race track is important. The image the fans at the track or on TV have is one consideration, but your sponsors may have another. Teams give a great deal of attention to the image they portray to their sponsors. Many Indy and NASCAR teams have one or more motorcoaches in which to entertain sponsors, potential sponsors, and VIPs, and conduct business at the track. These buses contain kitchens, wet bars, computers, fax machines, and large, comfortable areas for conducting business. They serve such diverse functions as café, lounge, private meeting room, and high-tech communications center, and the interiors are both functional and luxurious. Inside the coach, serious business is conducted and entry is restricted to a select few. Outside, under a large canopy, is an informal hospitality area for sponsors, guests, and team members. Dining tables, regular meals, closed-circuit television, and all-day snack buffets are normal features of professional motorsports entertainment. Some drivers even have their own motorcoaches for these entertainment benefits. If you are running a smaller team, the same approaches to sponsor and VIP entertainment can be used, even if the facilities are more modest. An RV can be used in place of a motorcoach, and a laptop computer and cellular phone can substitute for the communications center.

Between races, lunches and meetings with the sponsor may occur frequently. These will usually be used to plan events at the next race or promotional events, or to go over the most recent events. Sometimes your sponsor may want you to meet with a particular client at a separate event or party. These meetings and social events may sometimes seem to get in the way of the work you need to do as a racer. If this occurs, *DO NOT* shirk your responsibilities to your sponsor. Hire someone, if you need to, to do your work for you so that you can be at the sponsor's event. Be sure, whatever the cost, that you can meet your responsibilities to him. The cost of meeting these social obligations should be included in your proposal.

A Proposal for XYZ Western Wear

For a long time, consumer oriented businesses have relied on two choices with regard to their advertising approaches: print and electronic media. In the case of industries in which the target market is a large percentage of the general population, such as soft drinks and fast food, the print and electronic media can be very effective. If your target market is a small percentage of the population, though, this approach can be expensive and ultimately wasteful since many people hit in this "shotgun" approach may not be interested in your products. What is needed is advertising that hits your target market without wasting your money on those who are not likely to buy.

If the problem of aiming at and hitting only the target market could be solved, the advertising fat could be trimmed from your budget and your message made more effective. We have a method available to do just that.

Taking Aim

In order to design an ad campaign specifically for your business without wasting your money through the mass media, we must look first at the activities in which your target market are involved to learn where to reach them. High concentrations of Western Wear can be found in at least three different types of events: Rodeos, Country and Western clubs, and Auto Races, specifically at oval tracks. While you may already be using promotions through the first two of these, it is our understanding that you are not currently using Auto Racing as an ad medium. Please take a moment for us to show you how it can be successfully used to increase your profits.

Making Impact

Auto Racing is unique as an advertising medium first because of its **impact**. While radio, TV, and print ads may reach a large number of people and may even reach your target market, they are ineffective if they are not remembered. In order to be remembered, the audience must be bombarded with these ads so that repetition will create awareness. Ad people like to talk about the number of impressions created. Those impressions are not valid, though, if they are not ***lasting* impressions**. The excitement of Auto Racing creates vivid impressions of the sponsor's message each time race cars are seen either on display or on the track and those impressions last. It is not at all uncommon to hear fans refer to the "Folger's Car" or the "McDonald's Car."

Exposure

One of the most popular and fastest growing of all classes is for Mini Sprint cars. They resemble Winged Sprint cars, but are the size of Midgets and use 1200cc motorcycle engines. The races in this area are sanctioned by the Texas Mini Sprint Association and are run at numerous tracks throughout the Dallas/Fort Worth area on consecutive weekends. Although seating capacity varies somewhat among tracks, it is estimated that some **4000 people from your target market** will watch these cars **each weekend!** And because auto-racing results from local tracks are printed in newspapers, ongoing press coverage can be used to generate free exposure for the sponsor.

Regional level racing can also be used to generate exposure in nationally distributed enthusiasts' magazines such as *Open Wheel* and *Circle Track*. Race reports and feature articles on the cars are commonly accompanied by full color photos. This can be impressive to local fans and customers and can also be used to broaden your customer base into other areas where expansion is planned.

Even more exposure can be generated by displaying the cars at your locations for special promotions. Again, the impact to the customer of being close enough to touch the cars that they normally see on the race tracks will create **lasting *buying* impressions**. Combining Auto Racing with your other ad media, such as displaying the cars at Rodeos, is another effective method of creating real impressions with your target market.

Rifle Shot

For those special customers to whom you make large or frequent sales, Auto Racing can be very effective, indeed. Hospitality areas are frequently set up in which these important clients can be entertained and be made to feel like "insiders" at the racing events. By entertaining your clients in this way, they not only remember the excitement of the event, but also remember who made it possible—**you!** Building customer loyalty in this way builds sales for years to come.

Special promotions can be used to generate interest in your products among your customers by giving away a "Night at the Races," in which winners are treated to the same hospitality as your more important clients which, of course, makes loyal customers of them as well. Additionally, these nights at the races can be used as bonuses for your salespeople who have excelled in sales campaigns. Corporate entertainment can be used in a variety of ways through Auto Racing.

Make it Happen

These are some of the means by which we can help you to increase your sales by using Auto Racing as your primary ad medium. A specific program should be established which will best suit your needs and achieve your goals. We are in a unique position to help you do this for two reasons. First, having had experience in professional racing, we have been a party to the methods used by the top professional teams to generate sales for their sponsors at the highest level of motorsports. Second, we have close association with an advertising professional with over ten years experience in the business who can help you fine tune your campaign to generate the highest sales. Based on our experience, we are certain you will feel your involvement in Auto Racing will bring you the results you expect and need from your advertising budget.

The cost of advertising is high. But the cost of ineffective advertising is far more than just the dollars you spend. It results in lost sales, slow growth, and low profits. An **effective campaign** does not cost, it pays. It **pays dividends in increased sales, faster growth, and higher profits.** And with these come intangibles such as a higher profile and improved employee morale.

We are ready now to supplement your campaign for 1995 and preplan for 1996. The racing season is almost underway, and it is important to start soon in order to maximize your impact and exposure. We are eager to work with you on this project and help you reach your sales goals. Our representative will arrange a meeting a few days after your receive this proposal to answer any questions you may have and help you design an effective program.

Financial Requirements

Four basic packages are available and each can be modified to meet the Sponsor's special needs. Economies of scale allow us to do more with two cars, at less than twice the cost as shown in Packages Two through Four. Ranked in terms of cost and exposure, they are as follows:

Package One

One car is painted in the Sponsor's colors, and his logo is prominently displayed as the primary sponsor. The car is run on a local basis for the 1996 season, and is available for promotional displays for two weekend days per month through December. Included are writeups in area newspapers, race reports in newspapers, and radio where applicable. Hospitality tents can be set up at the race tracks when required at extra cost.

Package One cost $7,500

Package Two

In this package, two cars are painted in the Sponsor's colors and displaying his logo for the 1996 season, just as for the one car in Package One. Cars are available for promotional displays for two weekend days each month. Two cars deliver the impact of the Sponsor running a team instead of only a car. Publicity and exposure are generated as in Package One.

Package Two cost $13,500

Package Three

Package three consists of the exposure generated through Package Two by using two cars during the 1996 season, but is a better value by including hospitality tents and more promotional days. By using two cars, the effect on the consumer is more than the sum of the parts. Running two cars makes your operation look much larger as it is viewed as a team like the big time professional teams rather than a single car. Enhancing this image is the team's trailer painted in the sponsor's colors and displaying his logo. The cars will be made available for special promotions four weekend days per month rather than two as in the other packages, and Hospitality Tents will be provided at four events in 1996. Additionally, brochures will be printed and distributed emphasizing the sponsor's involvement in motorsports. All other media exposure is, of course, included.

Package Three cost $20,000

Package Four

This package is for the Sponsor who wants it all. Package Four is identical to Package Three except that no associate Sponsors will be accepted. No logos besides the Sponsor's will be displayed on the car except those decals required by the Sanctioning body.

Package Four Cost $25,000

Associate Sponsors

Although the primary sponsor will have the cars painted in his colors and have his logo most prominently displayed, provisions are available for those sponsors who wish to participate on a smaller scale. These sponsors' logos will be displayed in visible locations on the car, and photos will be available to the sponsor for display in his places of business and for customer handouts. Associate sponsors will only be signed after the primary sponsor.

Display location: Front Wing $750
Hood, both sides $1000
Tail, both sides $1000

Prices for Associate Sponsors are per car.

If your business dictates special requirements, programs can be tailored to meet your special needs. Costs and exposure can be modified as required. Please let us know how we can help you in this area.

Part Three: Going Pro

Chapter 10

Working with a Team

Opportunity usually knocks disguised as someone you know.

Getting Amateur Rides

When a driver drives a car he does not own, it is called "getting a ride." When this happens in amateur racing, it is most often a result of networking at the track. As the saying goes, "It's not what you know, it's who you know." The reality is it's both, but that's another story. There are several reasons that a racer would want to drive a car he did not own. Perhaps someone he knows has a more competitive car; perhaps he wants to move up to a faster class; or perhaps a friend has a car, but no engine, and the driver has an engine, but no car. There are endless variations of such reasons. The same can be said for the reasons that an owner might want to let someone else drive his car. I once got a ride because a driver friend got married and his new wife would not let him race anymore. Imagine my reaction when he offered to let me drive in his place! (It is widely known that racing is better than sex. Perhaps he disagreed!)

Amateur rides always come as a result of talking to people, so it is important to get to know your competitors as well as those in other classes. Opportunity usually knocks disguised as someone you know. One driver I know, who was building a new car, raced a car belonging to a competitor before his own was finished. The car owner had just begun a new job in a new city and could not race. I once made a deal to build a car for a friend and, as a term of that deal, arranged to be the driver on its completion. Getting a ride in an amateur class, like so many other things in racing, is a matter of being in the right place at the right time, so it is to your advantage to get to know many people and find out what they all are doing. In this way, you can maximize your chances of being there when an opportunity comes up.

When you do get an amateur ride, it is wise to remember that you are using someone else's equipment. Most club racers are on tight budgets, and a car owner will not be happy if he feels that you are abusing his equipment. Take care of it and treat it as if it were your own car. This does not mean you should not try to win, but it does mean you should be gentle with the car and not subject it to unnecessary stress. Keeping the car owner happy will ensure a long and pleasant relationship with him.

Getting One-Time Pro Rides

Of the two different ways a driver can get a pro ride for a single race, the most common is for him to rent a ride. This is most often the case when an amateur is trying to break into the pro ranks. In this case, either he pays to rent the car or he brings a sponsor with him to do it. What he gets for this varies with the team and the circumstances, but generally includes some testing or practice time before the race for which the car was rented, possibly some coaching by the usual driver or car owner, and, in endurance racing, a lot of track time.

When a contract is being negotiated to rent a car, the driver should make sure the following points are well understood:

1. **Limitation of liability.** Damage to the car may be the responsibility of either the driver or the team, depending on how the contract is arranged. Engine damage should only be the driver's responsibility in the event of a significant overrev. The driver should assume no responsibility whatsoever for damage or injury to another party arising from any incident.

2. **Amount of the contract.** The price the driver pays to rent the car should be clear, and penalties for damage to the car, or discounts for the car's inability to finish the race should be spelled out.

3. **Distribution of prize money and contingencies.** The split between the driver and team owner of any money won should be established. Contingency prizes provided by sponsors are made many times as a condition of finishing position. It should be understood who gets these and whether or not a split is expected.

4. **Track time.** A minimum amount of track time prior to and during the race should be established and discounts offered if this minimum is not reached.

A rental contract should be a written document, not just a verbal agreement and a handshake. There is no need for a formal contract drawn up by an attorney with all of his legalese, but an agreement should be typed up and signed by both parties. When it is all down in writing, the possibility of misunderstanding is greatly reduced.

The other way a driver can get a pro ride for a single race is a little different from renting it, but amounts to the same thing. When a pro driver carries personal sponsors with him, if he is any good at all, he is very attractive to a team who is in need of a temporary driver. Their need may arise because their regular driver is injured or because of scheduling or sponsorship conflicts. In this case, a driver may actually be approached by a team to run a given race. The team then expects funds from the driver's sponsor to pay their expenses.

When a driver gets a ride in this way, it is as if he were a paid driver. He assumes no responsibility for damage to the equipment, but his paycheck comes either from his share of the winnings or from his sponsor. He should not expect to receive a check from the team owner. The sponsor pays the team to let the driver race, but the team does not pay the driver.

This is a great way for a driver to work with several pro teams, get to know them, and let them see his professionalism, dedication, and desire to win. Many season contracts have resulted from such one-time drives.

Another method is sometimes encountered which is really a combination of these methods, but subtly disguised. Under the "Opportunities" classified ad columns in the back of racing magazines, ads are often seen which begin, "Sponsored Drivers Wanted." These ads are usually run by entry-level pro teams as a way to find someone to rent their car and pay the bills, but that is not to say that they are not legitimate teams. Some are competitive teams and help to pay for running their own car by renting out another. They look for sponsored drivers because they know that such drivers will have the funding to rent the ride. Their plan contains a flaw, however: Entry-level drivers do not usually have personal sponsors (unless it is Dad). So, these teams will take anyone, sponsored or not, who can afford their program. If you are considering renting a ride from one of these teams, check them out carefully by talking to their competitors.

Season Contracts

Pro drivers move from team to team frequently, something that any enthusiast who reads the racing magazines knows. Because of this, the time from a couple of races before the end of the season until about Christmas is know as the "Silly Season." It seems that the grass is always greener on the other side of the paddock, and drivers are sometimes willing to try another team in hopes that they can provide a better car, better engineering, or better funding. Trying to decide which move to make and when to make it is a pro driver's most haunting nightmare. In the '60s, Chris Amon was known as a very talented and proficient driver who never really gained the stature in racing he deserved because he always seemed to make the wrong decision when it came time to sign with a team. More recently, we have seen Alain Prost drop out of racing when he could not sign with the team he wanted, and Nigel Mansel move to Indy Cars and then back to F/1 for similar reasons.

At the professional level, a driver must be constantly be aware of the things going on within his team as well as the other teams. He must know, for example, which cars other teams will be using next season; what problems the engine manufacturers have had with particular powerplants; what the manufacturers are planning to do to overcome those problems; what new developments have occurred in tire technology; and which teams are tied in to which brands of tires for the next season. All of these things must be considered by a driver in choosing which team to join. In essence, he is planning a championship for the following season before the current season is finished, and is doing so based on what he thinks *might* happen. Every driver who signs with a team thinks he has made the best decision to maximize his chance of success. But, with so little information on hand on which to base their decisions, it is little wonder that so many drivers want to make a change at a season's close. Who could have guessed that the Penske organization would fail to qualify for Indy in '95 after dominating the '94 season so completely? Their drivers, Al Unser, Jr. and Emmerson Fittipaldi did not jump ship, but the events of May must have come to mind when it was time to renew their contracts.

Once a driver weighs all of the criteria at his disposal and decides which teams would benefit him the most, contract negotiations begin. Of course, a driver new to a series may have to beg for anything he can get. Winning the Formula Atlantic or Indy Lights championship will probably not get you an offer from Carl Haas. Assuming that his goals are realistic, though, and that the driver does have serious talks with a team manager about driving for a team, two things will be important. First is the amount of sponsorship money the driver can bring to the table. It usually takes several years for a driver to develop the business relationships he needs to carry a personal sponsor with him, but at the highest levels of motorsports, that's what it takes. The stronger you are in this area, the easier it is to get a ride.

Also important is the team manager's *perception* of how competitive the driver is. The word "perception" is emphasized here because it makes no difference how fast he is as a driver if, for example, he has not looked that fast previously because of a poorly set up car, an engine down on power, etc. For this reason, it is crucial for a driver early in his career to drive good cars and put himself in a position to always look good. Crashes, cars a little off the pace, and poor preparation will take their toll even on the best drivers, so it is important to minimize their effects. Team owners and managers read the racing magazines like the rest of us, and they watch races other than those in which their cars are competing. They do this to monitor the progress and development of up-and-coming drivers who may be useful to them a season or more down the line.

A minor point that may be of interest to the team manager is the driver's appearance on camera. This goes hand in hand with his perception of the driver. Young Tom Cruise look-alikes are more promotable than older drivers with the weathered faces of sea captains, all other things being equal. Fortunately, for all of us who are not movie stars, all other things are not equal. A driver's on-camera presence is considered, though, so he should practice giving interviews. If he does well on camera, it can be a plus toward renewing his contract for another year and keeping (or getting) sponsors.

As his career develops, a driver should also develop a résumé. Although team owners and managers will probably be aware of the driver's more memorable successes, the further information provided in the résumé may interest and impress them. Some drivers on the way up even employ agents or publicists to ensure that they get the publicity they need, and that the owners and managers of Indy Car, Formula One, and Winston Cup teams are aware of their successes. When used years in advance of the opportunity to drive for one of these teams, such tactics can be significant career-builders.

Another tactic that every driver at every level should employ is networking. As mentioned previously, opportunities arise as a result of being in the right place at the right time. Young drivers should cultivate relationships with those they come in contact with who can help them proceed in the future. Having a good relationship with a team owner or manager may, at some point in time, get you a test with the team. Frequently tests turn into testing contracts. When the regular driver is out for any reason, the test driver then subs for a race or two. And this may produce a season contract at a later date.

In professional motorsports, being an outstanding driver is not enough. A driver must also be outstanding at selling himself. The salesman in him should create opportunities that the driver in him can take advantage of. If he is a better driver than salesman, he should employ an agent to help make the opportunities happen, but even this does not totally relieve him of the responsibility of selling himself.

A driver must sell himself at every occasion.

Your Position in an Amateur Team

If you run your own amateur team, working in it may be simpler than being part of a professional team, but not much. In a team, each member plays a specific part and has certain duties he must perform. Although it is good for each member of a team to be *able* to perform in any position, it is not advisable to expect each of them to occupy every position. The same goes for you, the driver or team owner. In most amateur teams and many entry-level pro teams, the driver owns the car and most of the support equipment, and is therefore in charge. In this capacity, he must not forget that each team member has been selected to perform specific duties. If the team leader tries to do each team member's job as well as his own, friction will result, and the entire team will be less efficient. This problem is much too common. As in business, a team leader must delegate authority if the team is to be successful.

In a small team, each member may be required to perform several functions. The driver, for example, may also be required to make travel accommodations and build the engines. The crew chief may double as the engineer. Every team is different, but combining positions is the rule for amateur teams. Regardless of how these jobs are combined, it is important for each team member to feel that he has the responsibility for a portion of the operation. Without this responsibility, he may feel that he is dead weight or that the team lacks organization, and he may be right.

Your Position in a Pro Team

If you do not own the car (or team), but have negotiated a deal to drive for someone else, you should have a clear understanding of exactly how you fit in. Unless you are renting the ride, you should realize that the team will feel like they should be receiving something for your sitting behind the wheel. This something may be equipment you can provide which enhances the operation of the team, such as a trailer, or it may be something intangible, such as engineering expertise. It may also be the sponsors you bring with you, who can provide an increased operating budget, or simply the feeling that they can win more races with you, the world's best driver, at the controls. Most likely, however, you have gotten the deal because you can provide a combination of these. You are, perhaps, a good driver and can bring something else to the team besides your abilities.

Regardless of how you came to the team, you should realize that you are not the man in charge (unless you have bought the team, in which case you should read Chapter 11). As "just the driver," you are performing a specific function within the team, which is to drive the car. The other team members will have their own methods and procedures which have worked for them in the past, and will understandably be resistant to changes suggested by the "newcomer." You must adapt to their methods; do not expect them to change their methods to suit you. If their methods have not been working for them, then it may have been a mistake for you to join the team. The obvious exception is when the team has everything else necessary to win, but lacks the one non-material thing that only you can provide, such as superb driving or extraordinary engineering. If this is the case, they will probably be hanging on your every word.

If you have rented the ride, as is very common among all of the pro ranks, be aware that the only thing you have contributed is money. A rental driver is in a somewhat precarious position because he is temporary and the crew knows it. They are usually less sympathetic to the rental driver's problems and, although they may do a fine job of preparing the car, they may be less likely to go the extra mile for him. The rental driver's funds may be necessary for the team to race, but he is paying for the privilege of driving their car. The attitude of viewing the rental driver as the "temp" is especially prevalent if the owner of the team races a car out of the same shop. In this case, many rental drivers bring in their own crew to prepare the car. They can receive more personalized attention this way. As a rental driver, if you feel that your money has not been spent wisely, have a private conversation with the team manager. Hand waving and loud accusations in the paddock will not accomplish anything except to create animosity. The best bet, if you are considering renting a ride, is to go to a couple of races where the team is competing and watch how they go about the job of winning a race. If they look like a "Chinese fire drill," another team might be a better choice.

Some drivers who have had some success in club racing develop the attitude that they should be paid to drive or, at the least, receive a "fully funded ride." In reality, hardly any drivers are paid to race except at the highest levels of motorsports. Even at levels as high as Formula Atlantic, ARCA, or Trans Am, team managers will only consider giving a driver a "free ride" if he brings a substantial sponsorship package to the table. Most drivers club race for years before they get that elusive "fully funded ride," and even then they must find sponsors to pay the bill for them. That still does not earn them salaries, however. A driver may receive a portion of the winnings and may also receive contingencies from other sponsors, but he will not receive a check from the team manager. The contracts negotiated by Michael, Sterling, or Jacques will, most likely, contain provisions for paying the drivers. Until you reach this level, however, expect to pay your way whether with your own funds or with sponsor dollars. Sorry about that. That's just the way it works.

Working with a Team Manager

By the time most drivers arrive in pro racing, they have a considerable amount of experience. Many have years of club racing under their belts, and through their successes have accumulated the sponsors necessary to move up. Some of the youngsters have raced Karts or entry-level pro cars for several years and have good financial backing, usually from Dad. Either way, they are ready for professional competition and have money to contribute to a team or the sponsors who will. At this point in their careers, some changes will be made and include much more than just receiving a check rather than a trophy.

The adults who have been club racers have generally run their own programs and have gone about racing in their own ways. When they find themselves in F/2000 or Formula Atlantic as part of an established team that has been competing in the series for a number of years, they are likely to feel overwhelmed, not only by the amount of money they are paying to race, but by everything from the intensity of preparation of the car to the ferocity of competition. It usually takes a few races for a newcomer to adjust to his new surroundings.

The 21-year-olds, who will make careers of racing if they prove themselves good enough, are usually just happy to be in pro racing. They adapt quite quickly and quite well to their environment. It is their fathers who don't. More than one promising youngster has had a motorsports career ruined by a father who wanted to control every facet of his son's career, including every action of the crew. One Indy Lights team manager, speaking of a promising young lad who was driving for him, told me, "He would be fine if his father would just stay at home." The son eventually went on to bigger and better things, but that makes him the exception rather than the rule. Most of these kids' fathers are such a pain that the team managers find difficulty in achieving the team's and driver's full potential. When they encounter this type of situation, the managers will fulfill the obligations of the contract and look for a new driver for the following year. If you are the father of a young driver and would like to see him at Indy or in Winston Cup or F/1, take heed: The professional team managers know what they are doing. They have been doing it much longer than you. Leave them alone and your son will progress faster.

The team manager's job is to run the team and its multitude of functions. You, as the driver, are a small but important part of his duties. The team manager is your boss. You both have the common goal of winning races. If he feels you are charging too hard and taking too many risks which reduce your chances of finishing (you must finish before you can win), he will tell you to slow down, and you are obligated to follow his orders. If he gives you this instruction during a race, by pit board or by radio through the crew chief, it may be as a result of something you know nothing about, such as your nearest competitor having a problem. Perhaps he can see from your lap times that, although you are charging hard, you are slowing down, implying that your tires are going away. By doing the math, he may calculate that you can still win if you quit trying so hard. The driver has no time to analyze such information properly. As much as we sometimes like to think that this is a sport in which we can compete one-on-one with other drivers, it is still a group enterprise—a team sport. Recognize that the Team Manager may have information that you do not, on which he bases his decisions. This is why he is your boss. Follow his orders both on and off the track and your racing will go more smoothly and you will be more successful.

The relationship with your team manager is important. It must be based on trust and mutual respect if your races are to run efficiently. The Canadian Formula Atlantic standout, Patrick Carpentier, describes his relationship with Steve Cameron, his team manager, like this: "[W]hen Steve and I work together, we have complete communication and trust. That's a rare thing between a driver and team manager, but you can't realistically plan a championship without it."

Some drivers make the mistake of thinking that they are in charge because they are paying the team manager to race. This is always a mistake and never results in a good relationship or good results at the end of the season. If you want to win, give up the notion that you are the boss.

Working with an Engineer

In case you missed Chapter 8, Chassis Setup, I will mention again that the engineer has become an indispensable team member. *EVERY* team from Formula Atlantic and up has a full-time engineer who works closely with the driver on car setup. There is a lesson there for the teams running other classes: If you don't have an engineer, get one.

In larger teams, the engineering positions are broken down into specific but related specialties. People are required to run the data acquisition and telemetry systems and to analyze the data they produce; engine management specialists are needed; and someone is needed to interface with the driver to get his opinions of "what is going on out there." Of these, we will concern ourselves with the person in charge of the electronic hardware, called the data acquisition man, the person who interprets the data, and the man who works with the driver. In smaller teams, one person fills all three of these positions and is called the engineer.

To work well together, the driver and the engineer must have a special relationship. This relationship includes perfect technical communication. You as a driver must be able to explain to your engineer, in minute detail, what the car is doing. You must also explain it in terms he can understand. As you are developing your relationship, it may take a few races together for you to understand the questions he asks, and for him to interpret your answers into the technical terms he can understand. If you have some understanding of vehicle dynamics (even if it is not as deep as his) and he has driven cars competitively before (even if he is not as fast as you), the relationship builds much more quickly. He must know what you are going through on the track, and you must have some understanding of why he asks the questions he does. Together you must analyze what the car is doing and develop a theoretical priority list of the things that may improve it.

When cornering close to the limit, very small changes can sometimes produce large effects. This is due to the amount of weight on each tire at a given point in time. As an example, a minor change of the angle of one front wing can put an additional 10 pounds on the one corner of the car.

This may be just enough to bring the tire characteristics down the slope of the friction coefficient curve sufficiently to reduce understeer and balance the car. This is admittedly an oversimplified example because many other factors contribute to the car's handling, and there may be other more efficient cures for a particular handling problem. The engineer's job, though, is to determine what changes to the car's aerodynamics or suspension will be the most efficient. Your job, as driver, is to give him as much pertinent information as you can to assist him in making those decisions.

To do that, you must be aware of the three phases of a turn which have been discussed previously: the entry, the mid part, and the exit. These are terms you should be familiar with from analyzing your driving, and they are of great importance in analyzing the car's handling, too. Your engineer will want to know how the car is acting and reacting, not only in each corner phase, but also in each part of each phase. Corner entry understeer, for example, can be felt at the initial turn in point, or it can continue for a substantial portion of the turn entry. Your engineer will also want to know how the car responds to bumps, slick spots on the track, etc. (how the car *acts*) in relation to how it responds to your control inputs (how it *reacts*). It should now be clear that being a sensitive driver (as well as having a good memory) is highly beneficial to working with an engineer.

Few drivers have learned enough about vehicle dynamics to be excellent engineers in their own right. Craig Taylor is the exception. I had the opportunity to work with Craig for a couple of years in F/2000. Being a former navy pilot and since involved in commercial aviation, Craig wants to know all of the theory behind a subject, and then is extremely meticulous in the details of its practical application. He has enough knowledge of the intricate interrelationships among a car's suspension and aerodynamic parts that, together with being a very sensitive driver, he can generally analyze a car's behavior and arrive at a means of improving it. He does this through subjective impressions and logical thought. As a driver, his engineering knowledge is exceptional.

During my first race with Craig, after an on-track session, he pulled up next to his trailer, but did not get out of the car. I walked up next to the car, kneeled down, and asked, "Well, what's it doing?" To my surprise, he said nothing—for about 30 seconds. His helmet was still on, visor up, and he sat motionless with eyes closed. After that 30-second eternity, he removed his helmet, got out of the car, and told me *EVERYTHING THE CAR DID!* He said, "On the second lap at the entry to Two, it felt a little loose, but I went in a bit hot, and was on the brakes really hard so that may just be brake bias. But in Turns Four and Five on Laps Three and Five, I felt the same thing, so we may have a little too much bar in the rear. Since those are all right-hand turns, though, perhaps we should look at left rear camber and the shock on that side. And then on Laps Four and Five in Turns Three and Seven, etc." While he sat in the car, he was replaying a mental tape of the session. He remembered every single nuance of the car's behavior for the entire session. This kind of information from a driver is an engineer's most pleasant dream. Craig did not really need an engineer. I like to think, though, that he learned something from having another engineer to bounce ideas off of. I certainly learned a few things from him, among which is what it is like to work with a driver who really knows his stuff.

Craig is very fast and I attribute his speed to attention to the details of the car's setup, along with a healthy application of his right foot. At the other end of the fast spectrum is the driver who is just as fast, but doesn't have a clue about what the car is doing, and will tell the engineer so. This type of driver will get 110% out of a car, but cannot tell the engineer even the most fundamental things about the car's behavior. He depends on seat-of-the-pants impressions, and will get everything out of a car that is possible, regardless of what is given to him. However, he will have precious little to say about what it is doing or how to improve it. I helped one such driver to win a championship, but it was a struggle for both of us. In the middle are most drivers who have at least a modicum of sensitivity and memory and can tell their engineers what they feel. You should consider it your responsibility to at least get to this middle ground so that you can communicate with your engineer. He must set up the car's handling to suit your driving style and get as much cornering power and speed out of it as is possible. Your job is to help him do that.

A discussion of racing engineers would not be complete without mentioning professional engineers who drive race cars. These are exceptionally intelligent men who are very good in their respective fields, be it mechanical, civil, electronic, or other types of engineering. They have not, however, been trained in vehicle dynamics or aerodynamics, which are very specialized fields. Some professional engineers who have come into racing, on hearing of a team position called "the engineer," have decided that it is within their capabilities. It has never turned out to be true. To qualify as an auto racing engineer, a person must be fluent in vehicle dynamics and low-speed aerodynamics. Some colleges and universities have programs for aerodynamicists, but teach little of the Reynolds numbers, which are of interest to us, and only recently have any colleges or universities begun to teach any Vehicle Dynamics. At the turn of the millennium, diplomas and alma maters are not qualifiers of a good engineer, though, in time that may change. Today, racing experience and successful seasons are his credentials.

Michihiro Asaka is Honda's CART Project Coordinator. After suffering early problems with their engine in '94, Honda was ready to pull out all the stops to win in '95. Scott Goodyear was selected as a driver for Steve Horne's Tasman Motorsports Group, and did some preseason testing for Honda and Tasman. About Scott's testing Asaka says, "We were very satisfied with his technical performance. He was always consistent. We were very glad to see his consistent speed, and his comments were very good to our engineers. He is very precise. His opinion and our data are a compliment to each other. His comments helped us a lot, and I believe they helped Firestone a lot as well."

Working with an engineer can improve your driving, too. It is common for many club racers who are accustomed to flailing around the track in an ill-handling machine to become better drivers when they get rides in cars that are better set up. This happens all the way through the learning curve. After a season's testing and racing with Pat Patrick, Scott Pruett had this to say: "Working with Jim McGee and Steve Newey and all the guys at Patrick Racing has taken me to a new level. Having the confidence in the car week in and week out; having a car that works well, that does the things I want it to do, has let me fine-tune and move to a new level with my driving."

If your team has data acquisition equipment, both you and your engineer will have the privilege of interpreting the data derived from it. He will be able to look at the graphs and suggest that the oversteer "here" is produced by shocks or springs that are too stiff, but you will have the opportunity to say that the shocks seem fine through the bumpy sections, and therefore, you think it may be the springs. Data acquisition gives us the ability to quantify the effects of suspension and aerodynamic changes. These, along with the subjective impressions from the driver, help in identifying the areas that are causing problems and are keeping the car/driver combination from achieving its ultimate potential. When you have the opportunity to use an engineer, take full advantage of him. He can make the car easier to drive while making more cornering power, and that will allow you to push the car harder and be faster.

Chapter 11

Running a Team

I still say you can have the eight best mechanics in the world today and have the worst race team. It's all teamwork, it's all chemistry and that's the only thing I know how to do well. After that, they do all the work.—Barry Green

Selecting Team Personnel

Even though the driver is usually the one in the spotlight and receives most of the credit (and blame), racing is very much a team sport. The people behind the scenes who do most of the work contribute so much collectively to the overall effort, and specifically to individual areas, that no one of them is unimportant. Actually, since every nut and bolt on a car can affect finishing position, not only is each person on the team indispensable, but each task that a team member performs could well mean the difference between winning and not finishing. So much work is involved in preparing a car, getting it to the track, and getting it on the track, that it is impossible for one person to do an adequate job of it alone. It takes a team. But this is more than simply a collection of individuals with the same goals. The members of a team must work together cohesively and harmoniously in achieving those goals.

A well-organized, smoothly run team enhances your racing effort in many ways. The quality and consistency of your efforts reflects the thought you put into it. Observers, fellow competitors, and, most importantly, prospective sponsors can tell when you've got your act together. The benefits of self esteem and prize money are self-evident. In order to have a smoothly run team, it is essential to have good people, who will work well together, performing the important functions. If you are the team leader, it is your privilege to select the people on your team and your responsibility to lead and manage them effectively for the common good.

As the team leader, like it or not, you must develop the same management skills as are necessary in a business environment. Actually, management skills may be even more important in racing, particularly if you are using a volunteer crew. In a business environment, the boss has hiring and firing authority, which automatically instills in the employees a sense of duty to do things the boss's way. If your crew is helping just because they want to, they can pack up and go home anytime they feel unappreciated or just aren't having fun. Always remember why they are helping. They enjoy meeting a challenge and contributing to the team's success. If any member feels that he is working toward the team's goals, but those efforts are being thwarted by mismanagement of the team, he is likely to find another activity to occupy his weekends.

What should you look for in a good crew member? First, he should not have heavy time commitments outside of the team; he should be available to help. A job that takes him out of town frequently, many family activities, a nagging wife, or anything that keeps him from helping with car preparation or attending races will prevent him from being of much help to the team. It will also create animosity among the other crew members who may feel that they are giving more than their share of time to the project.

A good crew member should also get really turned on by the challenge the team faces. Sometimes, it is difficult to determine if a prospect really wants to help the team win or merely enjoys being around fast cars. Perhaps he only wants to sit in the motel bar and bench race or impress the ladies with his avocation. Many long, hard hours are required to be successful. A good crewman will enjoy the satisfaction that comes from doing whatever is required to meet that challenge.

Those are the two prerequisites every crewman should have: time and spirit. If someone is endowed with those two qualities, he can probably be turned into a valuable team asset. Other qualities are important, too, but can usually be developed by some on-the-job training.

A crew member's level of technical understanding is critical to his position. If one of your crew members barely knows which end of the car the engine is in, he can be valuable in a position that does not require heavy technical knowledge. Put him in charge of wheels and tires. Show him what needs to be done to mount and balance them properly and keep them from leaking. Teach him about rain tires, compounds, and how to "read" tires. In this capacity, he is a valuable member of the team, and he has a job to do in which he can learn and progress to the next level, as his understanding becomes deeper. In a capacity such as this, he also relieves other team members of these duties, so that their time can be spent on the other things they are proficient at.

A crew member's level of technical skill is important, too. Perhaps your buddy, Joe, works in a paint and body shop, and can be of the most use to the team when put in charge of fiberglass, body repair, and paint. Another crew member may be an engineer by day, and can be put to work designing the components that will inevitably be needed to modify the car during the course of a season. Analyze each person's skills and put them where they can be of the most use. A team member who makes his living as a salesman might be better suited to getting sponsors than building engines.

Although making the best use of each member's skills is important, individual needs and desires should not be overlooked. Your crewman who is a professional engineer may want to get his hands dirty when he comes to work on the car. The paint and body man may be tired of working with Bondo all day and want to do something different in the evenings. A team member should want to see the team succeed, but each will have his own personal reasons for being involved, too. When recruiting members and sorting out duties, find out what these motives are.

A good tactic to use in assigning duties and responsibilities is to rotate positions. In a small team, it is important for each member to be comfortable with every job. Sooner or later, one member of the team will have to miss a race and the other members need to be able to fill in for him. Giving each person experience in each position is a good way to ensure that each is competent in all areas. After a rotation, they may work out for themselves the positions in which they are most comfortable.

While some people will need supervision and guidance, others are natural self-starters and will question existing methods to find better ones. These people usually can think on their feet, solve problems, and need little supervision. Such qualities may be evident to everyone, and they may all choose such a team member who has natural leadership abilities to be their crew chief. By giving the crew some say in the decision-making process, you can reinforce each member's commitment to making the team work.

After your team has been assembled, each member's strengths and weaknesses will become evident. As the team leader, it is your responsibility to maximize the qualities that make each person on your team desirable and minimize the qualities that keep him from being perfectly suited to the job. Some means of analyzing crew member's performance relative to their positions is required.

In business, a method that has proved to be useful in evaluating employee performance involves rating the employee according to six criteria or dimensions: clarity, commitment, standards, responsibility, recognition, and teamwork.

Clarity relates to a person's understanding of the team's goals. Does he understand exactly what the team is trying to achieve and what should be done to achieve it?

Commitment is the individual's continuing understanding that those goals can be met. Was he involved in setting the goals, and does he feel that the team's performance is measured against them? If the answers to these questions are "no," his commitment may falter.

Standards should be very important to each team member. They should be established for each part of each routine job performed by the crew. They should also be constantly revised (upward) as the level of expertise of each member increases. Every well-run team will have higher standards at the end of the season than at the beginning. Thus, this dimension might also be called standards improvement.

Responsibility is each team member's feeling of commitment to his personal duties and the areas of team organization or car preparation for which he is responsible.

Recognition relates to the team member's feeling that doing a good or exceptional job will be rewarded, more so than lack of performance will be criticized.

Teamwork is exemplified by a crewman's feeling of belonging to a cohesive group, the members of which trust one another, take pride in their work, and enjoy the satisfaction that comes from accomplishing their goals.

In a racing team of a half-dozen or so people, sophisticated analysis is inappropriate, but analysis of each individual crew member on the basis of these criteria may be very useful in determining the overall health of your team and in determining how it might be improved. Use your managerial and motivational skills to improve the performance of each of your crew members. When each member of your team is dedicated, happy, feels needed, and is competent in his area of responsibility, your chances of success are greatly increased. Racing is more fun when the team has this kind of chemistry, too.

No one is perfect, so don't judge your personnel by the standard of perfection. That does not mean that we are not striving for perfection, though. Imagine your team working only normal hours in the shop, having the car ready when it is unloaded, and dominating your competition during the weekend. That is racing perfection. Manage your team so that it is working toward these goals.

Positions within the Team

In a typical amateur or entry-level pro team several positions need to be filled. With a small crew, some positions may be combined so that one person assumes the duties of two or more positions. Higher-level pro teams with more team members have the luxury of breaking the duties down, with each person performing only a small segment of what a crewman on a smaller team would perform. The following are job descriptions of the positions that should be filled on any team. They are not in order of importance because each is of equal importance to the team. Not one of these can be eliminated if the team is to be successful.

Team Manager

The team manager has ultimate authority. Everyone else works for him. His responsibilities include making sure that the proper parts and materials are ordered and arrive on time, that each member of the crew does his job, that the travel accommodations for the crew are made, that facilities for entertaining the sponsors are ready for an event, and that all of the other responsibilities of running a team are met. The team manager may not necessarily complete all of these tasks personally, but should delegate jobs to other team members. His responsibility is to ensure that everything gets done. In order to accomplish this, he must know what needs to be done and have understanding of how to perform each task. Barry Green says, "It takes someone to operate these organizations that understands every little problem in the team. Derrick Walker can do any job that is done by any of his members and so can I. We understand and I think that is helping us succeed in Indy Car racing today."

Crew Chief

The crew chief has the responsibility of overseeing the crew working on the car. He should be intimately familiar with all of the technical aspects of the car, and should be able to perform any task that he asks a crew member to do. The crew chief is the ultimate mechanic working on the car. He is also in charge of logistics at the track and the race strategy concerning pit stops.

Timing and Scoring

Although the sanctioning body's officials do timing and scoring for the entire field, each team should have someone doing the same for themselves. During qualifying, you should track any other drivers who may go faster than yours, and during races, it is important to keep track of your car's position and other competitors who may be closing or pulling away. Relaying that information to the driver may mean a better finishing position.

Fabricator

Good fabricators are scarce in this country. When you find one, hang on to him. The fabie's job is to make all of the special bits on the car which make it easier to work on, lighter, faster, or more efficient. He is a machinist, welder, sheet-metal worker, fiberglass (or other composite) man, and has generous amounts of common sense and experience.

Publicist

A team's publicist is responsible for writing and sending out press releases on the team's activities and accomplishments, negotiating deals with racing publications for interviews with drivers or other personnel, and getting the team mentioned in local newspapers, radio, and TV. Above all, however, the publicist's job is to get publicity for the team's sponsors.

Engineer

The team's engineer is the person responsible for setting up the car at a race event. He must have good communication skills to interface with the driver to find out exactly what is "going on out there." He should have a thorough understanding of vehicle dynamics, and be familiar with the intricacies of the type of car the team is running. If the team has data acquisition equipment, the engineer is generally responsible for its operation unless the team is large and has someone dedicated to this job.

Mechanic

A mechanic generally should be prepared to disassemble and reassemble any part of the car. In larger teams, this position is broken down into the following specialties:

Engine Specialist

As the title implies, this person is in charge of taking care of the engine. In a small team this may require rebuilding engines between races; in larger teams it may only require removing and replacing engines built by a professional engine builder. The engine specialist ensures that the engine is ready to perform when it is required to do so. He changes and analyzes oil and filters, laps valves when necessary, and reads plugs and alters fuel mixture or computer controls as required.

Gearbox Specialist

The gearbox man takes care of the gearbox and clutch. He changes gears, ring and pinion, and adjusts differential operation when required. He also rebuilds and/or blueprints the gearbox between races and documents clutch condition anytime the gearbox is removed.

Suspension Specialist

The suspension man attends to the uprights, hubs, bearings, A-arms, rod ends, pushrods, bellcranks, shocks, springs, swaybars, etc. He interfaces with the engineer to implement changes to the suspension to alter the handling or cornering power of the car.

As you can see, even with only these basic positions several people are required to do an adequate job. If a team is to grow and be successful, all of functions of these basic positions must be performed. However, many club racing teams try to cut corners and neglect some of these functions, such as timing and scoring and publicity. The result is inevitably a good finishing position lost by an official timing and scoring error for which the team has no backup data, or a lack of publicity when the team has an impressive victory. Successful racing is much more than just bolting a car together and driving fast. The teams that choose to do without the people who perform these all-important basic functions are destined to run in the middle of the pack and never gain the exposure and sponsors they seek.

Motivation

During the course of a season when the crew is tired from many long nights of car prep and miles in the transporter, it can be very difficult to keep morale up and keep everyone motivated, especially if victories are scarce. Bobby Rahal explains, "It's very difficult to keep a good crew, to keep a program together when there's a lot of frustration and a tremendous amount of work put forth." Having team members who have positive attitudes and tend to look at the bright side of each situation makes this job much easier, but everyone is prone to get the blues when victories are infrequent and meeting the goals seems to be getting more difficult with each race.

One way to keep morale up and keep everyone motivated is to be well-organized. Use lists both at the track and in the shop so that every member knows what must be done and in what order. Also list parts or supplies that are needed to perform the various tasks so they can be on hand when required. Keep the work area clean, neat, and well lit. No one likes to work in a cluttered, dark workspace and these conditions decrease performance as well as morale.

Make sure that all travel accommodations are made well in advance, and that each team member knows when the car must be done, when the truck will leave, and when and how he, personally, will get to an event.

Steer the team toward looking at its accomplishments rather than its disappointments. If the goal is to win the championship, but it does not appear that this will happen, talk about how far you have come and how well you are positioned for next season. Mention the pole positions you have taken, races you have won or have had high placings in, or the victories that almost happened. Looking at the team's accomplishments in the shadow of disappointments will help team members keep their heads up and morale high.

Above all, though, make sure that you employ on the team positive people who always find good things to say. As the saying goes, "It only takes one bad apple to spoil the barrel." Positive people are fun to be around, and this helps to develop the "chemistry" that helps make a team really work. A few encouraging comments from the team leader do wonders, too.

Logistics

The team leader must be a master of planning. He must know what jobs must be performed on the car after each race, and know what tools, equipment, parts, and materials must be available to the mechanics to perform those tasks. He must ensure that the race entries are mailed in at the proper time, know how long it will take the crew to perform their jobs and how long the trip to the track will take, and arrange for accommodations at the event and along the way, if they are required. If testing is to be a part of the trip, facilities must be arranged for the crew to repair any damage to the car that may result from the test sessions prior to the event. These are only some of the things that make the logistics of running a racing team difficult. Let's look at some of these in more detail.

The crew chief must be intimately familiar with the car. He should be able to predict, with accuracy, how long each facet of preparing the car will take. Some crew chiefs are more proficient than others at predicting the unforeseen and at estimating accurate times for completion. The team manager must know the personality of his crew chief and temper the "estimated times of arrival" accordingly. The chief should also know what parts and specialized tools will be required, and should inform the team manager early enough that they will be on hand when needed. The person responsible for rebuilding the gearbox will know that a selection of differential shims will be required to get the backlash right, and the suspension man will know that after several races, it is time to replace the hub bearings. The crew chief should know these things from the mechanics and either order the items himself or relay the information to the team manager, depending on how the team is set up. Either way, someone should be in charge or take charge of having the tools, parts, and materials on hand at the right time to get the job done. Telling a sponsor that the car was not ready because the team did not have a part when it was needed is not likely to instill confidence in the sponsor that you know what you are doing. It does not do much for crew morale, either.

From a logistics standpoint, the tougher items to handle are the unexpected ones. If not handled quickly and efficiently, they can turn into major crises. When a gearbox input shaft fails a crack test, with no spare on hand, the closest one on the other side of the Atlantic, and the next race only eight days away, panic is imminent. Such near-panic situations require creativity and clear thinking. A good support network helps, too. After the crisis is disarmed, the team members involved should be reminded of the need to check things that are likely to cause problems early so as to avoid a crisis situation. The budget should include a spare of the part in question, too.

Don't be too hard on the crew unless this is a continuing problem. No one can possibly predict when each part on the car will fail, and some history of the life expectancy of the car's parts is needed to have an idea of replacement intervals. Unfortunately, the only way to generate this history is to race the car and note what failures occur when. If the individual members of your crew have experience with similar cars, they may have clues as to which are the troublesome parts. This can be a great help in finishing races in your first season together.

In addition to the logistical problems of having the tools, equipment, parts, and materials available for the crew to prepare the car for an event, there are the logistical problems of getting the car, its support equipment, and the crew to the race. For a small team, this may be nothing more than having hotel reservations made and having everyone show up at the same time to load the car and leave. For a larger team, it is more difficult.

Pro race events usually begin with trailers occupying the paddock on Wednesday before a race. That day is used to park rigs, set up awnings, and generally get things ready for Thursday when practice begins. On occasions when the pro team must travel thousands of miles from their base to the race, it is necessary for a portion of the crew to leave with the truck and trailer several days ahead of the rest of the crew, engineers, drivers, team owners, and other support personnel. This means that the team manager must take care of both the transport of the car and the transport of the rest of the team at some later time. He must know when the car must be ready, so that it can be loaded at the proper time to make it to the track during the day on Wednesday. He must also make sure that the other team personnel have travel accommodations, and that each knows what those accommodations are. All of this must be done well in advance so that no foul-ups occur which result in someone being stranded in an airport or being late to the track for some other reason. And implied in all of this is that the team members who leave later have jobs to do in the shop after the car leaves.

To accomplish these goals, it is important for the team manager to have a good relationship with a travel agent. This will take a load off of his shoulders and may save the team some money, too.

It is also important for the truck drivers to have team credit cards for fuel purchases, motels, and emergencies, and a little cash, too. They should have all permits and paper work in order so that they are not detained by a state trooper, and should have maps, event entry information, and credentials.

The personnel who are driving or flying to the race later should also have the appropriate credit cards and cash, and have itineraries of where they should be and when they should be there to ensure that they do not miss a flight or other transportation or rendezvous. They, too, should have entry information and credentials. Bringing so many people and so much equipment together in the same place and at the same time so far from home is not easy. Chances of foul-ups are high, so care must be exercised to avoid them.

Expenses

How much does it cost to go racing? As much as is available. Although it would appear that the differences in financial considerations between the average club racer and an Indy Car organization are very great, the reality is that each must operate within a specified budget. The difference in the amount of money available to each is significant, but the manner in which it should be spent is similar. It is true that the club racer must frequently make do with an existing part of questionable quality, while a well-funded team will replace a perfectly usable item at regular intervals just to be safe. In each case, though, the team's money must be budgeted to make sure that it does not run out too soon. For example, a team that does not budget its money may have the most competitive and well-prepared car halfway through the season, but no funds left to compete with. This has happened.

The cost of competitive racing goes up as a function of two things: power and traction. As the adage goes, "Speed costs money. How fast do you want to go?" If speed were the only criterion, however, the cars that run on the salt flats would be the most expensive in the world. In reality, the most expensive cars are the ones that produce lots of power and a great deal of traction. Most cars have a reasonable balance between the two. Be prepared, though: As either power or traction is increased, whether through modifications to the car or by changing cars or classes, running the car will be more expensive. As an example, I once designed and built new suspensions for a GT-3 car and expected greatly improved cornering power from them. In the car's first race, the oil pickup in the pan was uncovered in the turns resulting in a lunched engine. The owner then had to fit the car with a dry sump system to ensure that the problem did not recur. Thus, it is a good idea to allocate some extra funds for unexpected maintenance expenses and updates during the season.

In order to determine how much it will really cost to race for a specified period of time, it is necessary to budget your available funds, just as you budget your time. You will find that there are two main types of expenses in racing: car preparation/updates and travel. These can be further broken down into the following categories.

Preparation and Updates
Parts and Materials
Tools
Labor
Outside Labor
Overhead

Travel
Truck and Trailer Maintenance
Fees
Tires
Fuel
Hotels
Food

These categories can be modified, some can be deleted, or others can be added to suit specific race organizations.

The first step in planning a budget is to determine how much can and should be spent in each category. This takes a considerable amount of thought and planning in that the category in question must be analyzed in relation to each specific race. A three-day event, for example, will require rooms for the crew for at least two nights whereas a normal club race would only require one night. That same three-day event may wear out or break more parts than normal and may require an additional set of tires. All of these things must be considered to come to a realistic estimate of cost for each category and each race.

You may find that planning a budget will make you more aware of the expenses involved in campaigning a car, but you should also prepare for those unexpected costs. You may find yourself on the pole for a given race with tires not up to the task of winning. Few racers in this situation can stand the temptation to use fresh tires to start the race. You may also find that something like clutch life may be less than expected, requiring additional outlay during the season. To avoid taking money that has been budgeted for other items to pay for these, some successful racers have found it a good idea to multiply the original budget by 120% to 150%. Racing *ALWAYS* costs more than you expect.

Unexpected costs bring up the question of how to stay within a budget after it has been created. The first obstacle that gets in the way of the budget is changing goals. For the club racer, it is much too common to find that your personal life intrudes on racing. Your wife's vacation, for example, may coincide with a planned race, forcing you to reschedule the events you attend. Perhaps an out-of-division race must be run instead of one closer to home. For many amateur racers, personal lives come before racing. (After all, if the wife's not happy, no racing takes place!) The reality of racing, though, is that these changes always cost money.

Another example of a situation involving changing goals would be the team that starts the season expecting only to get enough points to go to the Runoffs™, but halfway through the season finds that they have a real chance of winning their division. Like the racer on the pole who wants fresh tires, this team will stop at almost nothing to reach their new goal, one that was not reflected in the original budget. The first tactic for staying within budget, then, is to withstand the temptation to modify your goals. If you can't, find an additional source of racing funds.

Another tactic for staying within budget is to do a bit of cost accounting. You need not be a CPA, but a spreadsheet can show where you are over budget and where you may have money left. Most racers avoid the paperwork associated with racing like the plague. It is much more fun to work on the car or plan performance improvements. Creating a budget spreadsheet prior to the season and comparing it with actual expenditures after each race will enable you to spot these trends early, and is also valuable information on which to base a budget for the following season. Detection of over budget expenditures two or three races into a season may allow you to spot trends early and give you enough time to make the necessary changes in the way the team works to avoid a catastrophe later in the season.

For larger teams, this method is often employed along with a purchase order system, in which the crew chief writes a purchase order for all but minor purchases, which is then approved by the team manager. If a team is running more than one car, this system allows tracking of expenditures for each car, which again allows the Manager to spot trends. Your new hero driver, for example, may be going through three times as many dog rings as his teammate, and the decision can then be made that a test day to teach him how to shift will ultimately save money during the season.

If you can show that your team is not a non-profit organization, keeping good financial records may mean that your racing expenses are deductible when April 15th rolls around. If you find that the accounting is not getting done, get someone outside of the team to do it; a CPA will usually do the books for a small team for $50 to $75 per month. This can be a very good investment that pays huge dividends at the season's end.

For the novice who may not have been around racing a great deal, who has not seen what expenses are involved and how money can be spared when necessary, the best advice is to seek help from a veteran who runs up front with the same type of car. The more experienced racer has been in the same position, and can usually offer advice on which parts to buy and how to maximize their life to save money. He may even sell perfectly good but used parts at considerable savings. This is very common with tires. Many front-running drivers will use a new set in only a couple of sessions and then sell them for half price. This can be a great savings to the racer on a limited budget.

Chapter 12

Planning a Championship

There is no easy side to this business. I have to work hard in a lot of areas to ensure success.—Barry Green

Attitude

Anyone can do anything if he believes it is possible.

Regardless of what aspect of racing is being discussed, it always seems to come back to attitude. Perhaps this is so because a competitor's attitude is the foundation from which he approaches the task facing him. Planning a championship is no exception. The preparation necessary to win a championship begins with believing that you can do it. In addition, the person in charge must be proficient in a number of areas, or must surround himself with people who are. Assuming that you have these bases covered, and have the right attitude, there should be nothing preventing you from achieving your goals—even if one of those goals is winning a season championship.

Improving Your Team's Chances

By now you should realize the value of both planning and organization in accomplishing complex goals. Each of these has been discussed before, but they are of such significance to a successful racing effort that they deserve another mention. Every aspect of your racing effort *must* be well-organized. The tools and equipment used in the preparation of the car should be organized in such a way that any crew member always knows where to find them. This reduces the time spent "chasing tails." The same goes for the parts and materials used on the car. The crew's duties should be planned so that they do not duplicate tasks, get in each other's way, or, even worse, end up in a situation where one member is waiting on another. Travel accommodations must be planned well in advance so that everyone makes it to the track at the proper times.

In addition to logistical planning and organization, the approach taken toward the goal of winning a championship is important. Two basic approaches are effective depending on how the championship is determined. In SCCA club racing, the results of only one race determine the National Champion. In most professional racing, though, points are accumulated by finishing position in each event. At the end of the season, he who has the most points is the Champion.

To win SCCA's Runoffs™, the first priority must be to qualify to enter it. SCCA requires a certain minimum finishing position within a driver's Division of Record before he receives a Runoffs™ invitation. In most divisions, this is not particularly difficult. If a driver believes that he will be competitive at the Runoffs™ level, he will certainly be competitive within his division. If this is the case, two things are still of vital importance: His car must be reliable and he must make the majority of the races to accumulate points.

When Dave Salls and I were working together, we got a late start in the season and were afraid there was too little time to get enough points to go to the Runoffs™. After winning the first two races, we were less apprehensive, but had an ignition failure while leading the third race. Dave was about 150 feet short of completing half of the race distance when he stopped in the pit lane. Feeling that fourth place points might turn the tide when it came Championship time, he cranked across the line on the starter, was listed as a finisher, and got those points. Our goal was to win the National Championship. To accomplish that, we had to obtain enough points to go. As it turned out, we won every other race except one in which Dave crashed and did have enough points to make the trip to Atlanta. It is ironic that he won the Runoffs™, but earlier in the year we were not even sure we would be able to go.

After accumulating sufficient points to go to such a race, the next priority is to be ready for it. Part of this involves analyzing your team's strengths and weaknesses. It is imperative to make the most of your strengths and arrange the race to minimize the areas in which you do not excel. You might, for example, compile a list as the one shown on the following page.

Putting the pros and cons of your team's chances of winning the race down in black and white helps to focus attention on the methods by which you may achieve your goals. In the scenario described on the next page, for example, you might want to try to drive a smooth, consistent race and pick off those in front one at a time while conserving tires. Since you may be down on power compared to some of your competition, but have better handling and cornering power, it would be wise to use your higher cornering speeds for closing on other cars and staying with them by drafting down the straight. This gives you the option of outbraking them at the end of the straight, since you will not have to slow quite as much for the turn. This is exactly what Dave Salls did at the Runoffs™ in 1991, and then pulled out a seven-second lead by the checkered flag. As it turned out, he still had plenty of rubber left, but had he not, he could have slowed a bit with that much lead and still gone on to the win.

One common mistake is made by dozens of Runoffs™ competitors each October. Some racers feel that to optimize their chances of winning or doing well in the Runoffs™, they should use all of

	STRENGTHS	WEAKNESSES
Preparation	Good reliability, overall prep has presented no problems.	Photocell on crank trigger is damage-prone. Inspect this area and install new photocell prior to race.
Cornering Power	Probably the best cornering power of any car in the field.	Trying softer tire compound. Possibility tires may be gone at end of race. No data acquisition system to help optimize car.
Handling	Car is well set up, Driver happy.	Weather may not hold. New setup will then be required.
Power	Straightaway speeds are good. Power is adequate.	Other cars have slightly more.
Gearing	Better suited to the course than most.	Gearing not perfect but changing gears requires pulling engine. Will do so if time permits.
Driving	Smooth and precise. Driver may help tire life.	Not prone to take risks when necessary.

the "tricks" they have been contemplating all year. This may mean only using the Hi Po cam or completely rebuilding the engine with new cam, pistons, rod ratio, etc. Some racers may put on the latest suspension just before the Runoffs™, or the new clutch and transmission, or the bigger brakes. These are *always* mistakes. More races have been lost by unproven hardware than by all other factors except preparation. Try that new technology in another race or, better yet, a test session before expecting it to perform flawlessly at the Runoffs™. You will find the cam too peaky, the engine unreliable, the suspension needing to be dialed in, the transmission selector forks misadjusted and not shifting smoothly, or the bias unable to be properly adjusted on the bigger brakes. It happens every year. If you feel that you will not be competitive without the new pieces, either test them before the race or don't go. Anything else is just folly.

In a specific important race, be it the Runoffs™, the Indy 500, or, to a lesser extent, Le Mans, it is possible within limits to pull out all the stops and go for a win. If your championship is determined at the end of the year by accumulated points, your approach must be a little different. Although many racers look down with disdain when it happens, some championships have actually been won by a champion who has not won a race. What it takes is consistent high-finishing positions. Winning races certainly gains more points and boosts the driver's ego and the crew's morale, so winning is always the goal. Hanging it out to make a pass for the lead, however, and falling off the road and getting only a DNF will generate few accolades if second place would have clinched the championship. This is where a driver must use his intellect rather than his emotions, and this is just as true off the track as on.

When going for points, the attitude of the driver and the entire crew must be just a little more conservative than if only one race is being considered. The reliable engine that makes a little less power may be the better choice than the "killer motor" that likes to tear itself apart. The engineer can set the car up to be a bit faster, but harder to drive. This is okay for qualifying or an all-or-nothing attempt, but the strategy for a season-long setup should be more conservative in order to give the driver a better chance of avoiding mistakes and therefore finishing races. For the long term, the mechanics may provide a little more airflow to the radiator, oil cooler, and brakes, knowing that the car may be more reliable but not quite as fast as a result.

Usually a series champion wins one or more races on the road to his championship, but has several seconds and thirds along the way. Other contenders for the title may win a couple of races, but those may be accompanied by a string of DNFs. Rarely ever will these later competitors come out on top. The keys to winning a championship based on points are:

1. Start all the races
2. Finish all the races
3. Win if you can
4. Place highly if you cannot
5. Never, never, never DNF!

Evaluating the Competition

As discussed in Chapter 7, part of winning is knowing your competition's capabilities. For a season-long championship, this is a little more difficult than for any given race. As the season progresses, you and your team are discovering shortcomings and correcting problems to improve your finishing positions. Your competition is doing the same. Properly evaluating the competition requires a great deal of awareness of the nature of the sport and the things that are happening within it. Just as drivers are forced to make judgements on the competitiveness of a team when making contract negotiations, each team owner/manager must continually assess the competitiveness of other teams based on the things he can find out about them or the things he discerns through instinct. Knowing that a team is having trouble with their engine reliability, for instance, could make one suspect that they could be outlasted in longer races. If, however, it is known that the person in charge of their engine development is regarded as a miracle-worker, and the next long race is two months away, your team's perceived advantage in reliability may turn out to be imaginary. Alternatively, your competitor's long-heralded new car unveiled at the first race may look really trick, but have teething problems lasting far into the season. Things are not always as they seem, and it is the racer's responsibility, if he is to be successful, to find out how they really are.

This is an area that requires good social skills in making friends with the competition and staying aware of all of the offhand remarks and small inflections of tone and body language among competitors in the paddock and at dinner on Saturday night which may prove meaningful. I am not trying to imply that you should cultivate friendships among your fellow racers just for purposes of espionage. I have made some very good and long-lasting friends of some of those I've raced with. The fact remains, though, that a great deal of useful information about a team's problems and triumphs is exchanged when the mechanics are having a beer.

Pro teams generally have a fair bit of trouble keeping any speed secrets they are able to devise simply because many of their mechanics know each other and socialize at race events. Even though it would be to each team's advantage for all of them to keep their mouths shut, it is human nature for each of us to want to talk about the things that are happening around us. When one mechanic comments that his team is having trouble with brake temperatures on their cars, another may say something like, "Yeah, we had that problem at Elkhart last year and what we did was ..." Even more technology transfer occurs when mechanics or other crew personnel change teams, either within a season or between seasons, and take the lessons learned at one team to another. Sharing of information is a fact of life for pro teams, even though they try hard to keep their setup secrets hidden. It is one of the primary reasons that technology accelerates so rapidly in racing, and it shows no signs of ever slowing down. To use this to your advantage, you and your entire crew should always be on the lookout for information on what the competition is doing, from whatever source it comes. At the same time, your crew should be acutely aware of what they say at all times. In another era the slogan was, "Loose lips sink ships." These days loose lips can sink your chances for a championship.

When information is gained about a competitor, it still must be the subject of intense scrutiny. Evaluate it the same way you would any other information. Is it accurate? Is it important? How can we take advantage of it?

Evaluating Your Team

Evaluating yourself is usually more difficult than sizing up your competition. We always tend to think that we can do no wrong, and make excuses for our mistakes, rather than confront them head-on, realistically and logically. Using a rational approach, though, can reveal things about your team which can mean the difference over the course of a season between coming out on top and being second best. You should evaluate yourself and your team in several areas: driving, car prep, race strategy, aerodynamics, power, and car setup.

Driving. A great deal of emotion is involved in a driver's perception of his own driving. His identity is usually somewhat caught up in how he thinks of his driving ability. When you evaluate your driver's performance, therefore, you can look at his lap time improvements at a certain track, compare data graphs from his best lap at this race to the last time your team was here or by any other comparison method, and he may still have an excuse for whichever was the poorer performance. For this reason alone, then, when striving to improve your driver's performance, it is wise to look at all of the data and examine all of the extraneous factors that may have influenced the data, before confronting him with it. These factors may include tires that had gone off, being in traffic, an engine that had poorer leakdown figures, a damp or oily track, or dozens of other factors which may each be excuses or legitimate reasons for a change in performance. In order to help isolate the driver's ego from other factors, it may even be beneficial to have meetings with him (as well as each crew member) after each race for the purpose of determining how his performance might be improved next time. When he knows that he is not being called on the carpet, but is being asked to help the team do better, his ego plays less of a part, and he is likely to be more cooperative and more honest about his driving and racing performance.

Evaluating your team's driver is even more difficult if you are at the wheel. If you feel that you may not be able to honestly and without prejudice assess your own driving, first solicit the opinions of your crew. Through watching you on the track, they may be able to tell you that you are not staying on the throttle as deep into Turn One at Atlanta as other drivers, or that you do not seem as willing as some others to take chances. It is also beneficial to consult well-respected drivers in other classes or perhaps instructors at a driving school to ask their opinions of your techniques. The goal is to constantly improve. It is important to take all of these opinions to heart so that you can become a better racing driver throughout the season and improve your chances of winning the championship.

Car Prep. Car preparation mainly affects finishing races. As the saying goes, "You can't win unless you finish." The crew members are often responsible during the season for making small improvements to the car which make it faster, too. This may be as complex as noticing that a U-joint problem is really a result of the rear shock action, or as simple as relocating the battery to obtain more favorable corner weights. All improvements of this kind are performance items, and it should be realized that the crew is able to improve lap times in addition to reliability. The frequency of these performance enhancements should be considered along with the reliability of the car when evaluating the performance of the crew. Being human (most of the time), mechanics will make mistakes. These will usually be minor ones which will cause a bit of extra work or may even cause a session to end prematurely. The more significant ones, however, are those that cause retirement from a race or end in a disqualification, or, worst of all, lead to a crash. When analyzing the crew's performance, as with the driver, be sure to look at the situation from all angles to get an accurate picture of what happened. Only then do you have enough information to determine how to improve the preparation of the car.

Race Strategy. The world can be a pretty unpredictable place. No matter how hard we try to gain an advantage through strategy something unexpected can happen and our best-laid plans fly right out the window. If your race strategy works in fifty percent of the races during a season, you are doing very well. As with all other activities in racing, practicing your strategy prior to each race will improve your success rate. Comparing the actual events of a race with your pre-race predictions will improve your awareness of the things that are important to the outcome, and will assist in improving your ability to accurately predict them.

Aerodynamics. This category can be further broken down into downforce and drag. The goal of the aerodynamics of modern race cars is to generate the most downforce with the least attendant drag. The name of the game is compromise. Downforce can always be increased, but drag will increase, too, and sometimes not proportionately. The balance point, the optimum downforce-to-drag ratio, is very elusive. If too much downforce is generated, the car will fly through the corners, but be slow down the straights. If too little downforce is generated, the car will be slow through and out of the corners, and, as a result, will be slow at the end of the straight, too. It is usually better to have too much downforce than not enough, although this condition, too, will be slower than optimum. Assuming that the straight is long enough that the maximum engine revolutions and gearing do not play a part, top speed is dictated by horsepower and aerodynamic drag.

Power. It would seem that the power of the engine would be easy to evaluate from the car's speed at the end of the straight. Once again, however, the answer to this question is not that simple. As we have seen, horsepower and aerodynamic drag are mainly responsible for a car's ultimate top speed. A higher speed entering the straight, though, will result in a higher speed at the end of the straight, providing the car is accelerating for that entire distance. A couple of exceptions are the long straights at Atlanta and Brainerd, where the top speed is reached prior to the end of the straight. At most tracks, the car's cornering power and handling and the driver's ability are all tied up in the car's speed at the end of the straight, and it is sometimes difficult to depend on speed to determine competitive horsepower. If you have competitive speed at the end of the straight, it is a sure sign that the engine is producing sufficient horsepower. But if your car's speed on the straight is low, it could be because of lack of horsepower, poor handling or cornering power, improper gear ratios, or improper driving technique.

Using dynamometer figures is almost as perilous. All dynos should read the same using corrected air temperature and pressure. They don't. Some engine builders, I am convinced, inflate their numbers to make their engines look better. Other discrepancies occur as a result of measurement irregularities. An engine producing 146 HP on one dyno may only make 141 on another. Unless back-to-back tests are made on two engines on the same dyno, it is impossible to tell which makes more power. It is also useless to try. Assuming that your driver has a competitive right foot, his car is reasonably well set up, the car does not have an aerodynamic problem, and you are getting competitive speed at the end of the straight, you have competitive horsepower. That is all that is required, and your time and money will usually be spent more wisely on other performance pursuits.

Car Setup. As racers continue to develop their cars through new technologies, the cars become ever more complicated, requiring even greater knowledge of setup. Whether the driver or an engineer is responsible for this important function within a team, a great deal of technical knowledge is essential. Books such as Smith's *Tune to Win* and Milliken's *Race Car Vehicle Dynamics,* as well as *Racecar Engineering* magazine are excellent sources of information for this purpose. Along with technical understanding, the individual responsible for setup must have a good deal of experience, too. Racetracks have a nasty habit of throwing situations at an engineer which he has never considered. Having related experiences, though, gives him a starting point from which to analyze a new situation and conceive a method of improving the car's performance in that situation. The only way most engineers can generate this experience is through time with the car on the track.

Developing a Season Strategy

At the beginning of each season you should have a plan for how you expect to win the championship. Rarely will that plan remain unchanged by the end of the season, but it will give you a blueprint that you can work from to create successful results. As an example of such a preseason blueprint, consider the following scenario:

You have put together a reasonably efficient and competitive stock car team and have been running together for the two previous seasons. You have finally secured a major sponsor, although the total amount of money injected into the team is not as much as you had hoped. During the winter, you had a new car built by a professional chassis builder, and you expect that the last year's handling problems have finally been solved. Even with your old car, you finished third in championship points last year. The two teams that finished ahead of you are established and consistently fast. In addition, a new team appeared mid-season last year, which you expect to be tough competition. The schedule includes mostly ½-mile paved tracks, but has a few one-mile races, too. These longer tracks are your weak point, but one of these races has a quite large purse and you are tempted to run it and try to finish well in order to supplement the budget.

The question then arises, "Do you try to win this one long-track race so as to be better able to afford the rest of the season, or do you concentrate on what you do best, the short tracks?" Obviously, going for the big prize at the long track is a gamble. To be in the best position to do well in

The Importance of Winning a Championship

In his autobiography, A.J. Foyt says that a driver should win a championship before going on to another class or division. A great body of evidence supports this view. Paul Tracy, Michael Andretti, Al Unser, Jr., Didier Thieys, Frank Freon, and a host of other professional drivers all subscribe to this school. Each has won a championship in a particular division before moving on to another. This is a way of proving to the world and anyone else paying attention that the driver is the best in his class and deserves to move up.

Every situation is different, though. Some of these very same drivers have moved up even after a less than perfect season. When Price Cobb began racing in a Formula Ford he said, after only three races, "This is boring. I want something faster!" Michael Andretti was never a Formula Ford National Champion, but moved on to Formula Atlantic where he won the Championship. Al, Jr. bagged a Can Am championship, but failed to win an IMSA title. Gilles Villeneuve was a standout in Formula Ford in Canada and the class of the field in Formula Atlantic, but gained a Formula One ride with Ferrari without a championship to his credit. Championships provide a driver with a great deal of credibility with fans, sponsors, and peers, but they are not a prerequisite to moving to a faster, more prestigious, or more competitive class.

The real prerequisites should be to believe that, even if you did not win, you *could have* done so, and to figure out what should have been done differently to achieve that goal. Such knowledge may come down to a specific event. For instance, a mechanic reinstalling a halfshaft backward may have cost you a finish, and with it, the championship. Or the it, you should budget for testing at the same track. By doing so, you may learn why long tracks are your weak point and solve the problem once and for all. This puts you in a better position for that one race and all future races on one-mile tracks. The testing required may use up whatever winnings you expect at that race, though, and your financial well-being may not be improved.

Alternatively, you could concentrate on the short-track races, in which you know you are competitive, and not even go to the others. Last year, you received no points for the one-mile races you ran and still finished well in points. Not attending the longer tracks may improve your finances, while not hurting your championship chances.

Another alternative would be to combine the best aspects of the other two. You have been running the past two seasons with help from whatever sponsor you could pick up at each race. Perhaps continuing to do so while keeping your new sponsor happy would give you the funding you need to do the long-track testing you need to do to be competitive at these events. You may need to

team that you picked (or that picked you) may not have been as proficient as another. Although making excuses for not winning is undoubtedly a mistake, rationally analyzing the factors that kept you from winning is a healthy thing. Careful consideration will show you how to refrain from making the same mistakes again, and will also reinforce your confidence that you can still win even if it did not happen this time.

Coming close to winning a specific championship repeatedly without moving on delays a driver's progress unnecessarily. When a driver proves to himself and the racing community that he is competitive, and that it is only a matter of time before he wins the title, it is probably time to move up to another class where there is more to learn.

Conversely, some drivers try to move up too quickly. They may run in the middle of the pack of a particular class for a season before moving to a faster and more challenging class, and then find themselves floundering around in the middle of the pack there, too. What such a driver needs is to learn how to win in a given class before moving up. By that time, a driver has learned most of the lessons that the class and car type can teach him and is ready for the new lessons of the faster class.

Having a championship or two on a résumé is certainly no bad thing. Spending years in the same class trying to obtain those bragging rights, though, has halted otherwise promising careers.

change your focus from how to make the best use of the resources you have, to how to improve those resources with the least risk of money and points. That would mean more work in finding those one off sponsors (work you had hoped to avoid this season), but it would also mean doing that long-track testing and solving a problem that has been holding you back.

Making these decisions prior to the start of a season is crucial to developing your blueprint. Once they have been made, you can focus on the one path in front of you that you have decided will most enhance your chances of achieving your goal. The other factors that have been discussed, such as the strong and weak points of your crew and the identity of your biggest competitor's new engineer, should also be considered in developing that blueprint. Try to outline it in detail, down to such things as practicing pit stops, if that is a weak area. During the season, you may discover a better way to reach your goal and modify your plan, but it is imperative to have a map to follow. Otherwise, it is much too easy to lose sight of the goal and make a costly decision.

Championships are designed to prove who is best—best on one day for championships decided by one race, and best over the course of a season for those determined by points. If your goal is to win the championship, then you must make up your mind to be the best under the prevailing circumstances. Determining whether or not your team can be the best is a matter of honestly assessing your team's capabilities along with those of your competitors. If your championship requires that you collect points during the year, you also have the opportunity to change those things you find to be detrimental to your effort, and those things you find another competitor is doing better than you are along the way. At each race, give thoughtful consideration to the reasons that another competitor finished higher than you did. Analyze the race from both your perspective and his, and determine the logical reason why he finished ahead of you. Leave out the emotional elements such as, "He has a bigger budget than we do and could afford two sets of tires at each race." Or "He cut the inside of the turn on Four and threw rocks at me. When I backed off, how was I to know I wouldn't catch him again?" These emotional responses cover the real reasons that your competitor finished ahead of you. In the first case, he did a better job of finding sponsors to fund his effort and could afford additional tires. In the second, he came up with an effective plan to get you off his rear and defend his position. Be honest with yourself, learn from your mistakes, and use the lessons you learn to improve your position in the next race. Continually improving is just as much a part of winning a championship as starting the season strong.

Appendix A

Driving Schools

Akin-White Endurance Racing School
4320 West Osborne Ave.
Tampa, FL 33614-6926

American Speed Racing School
Lakeland, FL
941/927-3827

Buck Baker Racing School
1613 Runnymede Lane
Charlotte, NC 28211
800/529-BUCK
http://www.buckbaker.com

Skip Barber Racing School
29 Brook Street
Lakeville, CT 06039
800/221-1131
http://www.skipbarber.com

Skip Barber West Racing School
P.O. Box 629
Carmel Valley, CA 93924

Bob Bondurant School of High Performance Driving
P.O. Box 51980
Phoenix, AZ 85076-1980
800/842-RACE
http://www.bondurant.com

Danny Collins Racing Schools
1626 Albion Street
Denver, CO
303/388-3875

The Driver's Connection
Willow Springs Raceway
P.O. Box X
Rosamond, CA 93560

Elf-Winfield
1409 S. Wilshire Drive
Minnetonka, MN 55343
612/541-9461

Formula 101
Harrisburg, PA
717/944-4368

Jim Hall II Kart Racing School
805/654-1329
http://www.jhrkartracing.com/

Pitarresi ProDrive Racing School
1940 N. Victory Blvd.
Portland, OR 97217
503/285-4449

Porsche Precision Driving School
P.O. Box 11912
Fort Lauderdale, FL 33339

Powell Motorsport Advanced Driving School
Rural Route 1
Blackstock, Ontario, Canada L0B 1BO

Road Atlanta Driver Training School
5300 Winder Way
Braselton, GA 30517

Bertil Roos Racing School
P.O. Box 221
Blakeslee, PA 18610
800/722-3669
http://www.racenow.com

Jim Russell Racing Drivers School
29305 Arnold Drive
Sonoma, CA 95476
http://www.jrrds.ukltd.com/

SCCA Enterprises National Racing School
14570 East Fremont Ave.
Englewood, CO 80112

Bill Scott Racing School
P.O. Box 190
Summit Point, WV 25446

Southard's School of Racing
P.O. Box 1810
New Smyrna Beach, FL 32168
800/422-9449
http://www.racingschool.net/

Spenard-David Racing School
Rural Route 2
Shannonville, Ontario, Canada K0K 3A0

Start Kart Racing School
Orlando, FL
800/782-2782

Stephens Brothers Racing
2232 S. Nogales Ave.
Tulsa, OK 74107-2826
918/583-1136
http://www.mavier.com/stephens_bros_racing/

Tracktime, Inc.
4464 Little John's Place
Youngstown, OH 44511

Ultra Karts Racing School
316 C Avenue
Coronado, CA 92118

Appendix B

Recommended Reading

Books

The Art and Science of Grand Prix Driving
Niki Lauda
Motorbooks International
Osceola, WI, 1977

Bob Bondurant on High Performance Driving
Bob Bondurant and John Blakemore
Motorbooks International
Osceola, WI, 1982

Data Power
Buddy Fey
Towery Publishing
Memphis, TN, 1993

Driving in Competition
Alan Johnson
W.W. Norton & Company
New York, NY, 1971

Engineer to Win
Carroll Smith
Motorbooks International
Osceola, WI

Grow Young with HGH
Dr. Ronald Klatz and Carol Kahn
Harper Collins Publishers, Inc.
New York, NY, 1997

How to Make Your Car Handle
Fred Puhn
HP Books
Tucson, AZ, 1976

Improve Your Vision Without Glasses or Contacts
Dr. Steven Bersford
Simon & Schuster
New York, NY, 1996

Nuts, Bolts, Fasteners and Plumbing Handbook
Carroll Smith
Motorbooks International
Osceola, WI, 1990

Prepare to Win
Carroll Smith
Aero Publishers
Fallbrook, CA, 1975

Race Car Engineering and Mechanics
Paul Van Valkenburgh
Published by the author
Seal Beach, CA, 1992

Race Car Vehicle Dynamics
William and Douglas Milliken
Society of Automotive Engineers
Warrendale, PA, 1995

Racer's Travel Guide
Judy Preston and Carroll Smith
Motorbooks International
Osceola, WI, 1991

Sports Car and Competition Driving
Paul Frere
Robert Bentley, Inc.
Cambridge, MA, 1966

Strength Training for Performance Driving
Mark Martin and John Comereski
Motorbooks International
Osceola, WI, 1994

The Technique of Motor Racing
Piero Taruffi
Robert Bentley, Inc.
Cambridge, MA, 1958

The Theory and Practice of High Speed Driving
Walter Honegger
Speed and Sports Publications
London, England, 1971

Think To Win
Don Alexander
Robert Bentley Publishers
Cambridge, MA, 1995

Tune to Win
Carroll Smith
Aero Publishers
Fallbrook, CA, 1978

A Twist of the Wrist
Keith Code
Acrobat Books
Los Angeles, CA, 1983

A Twist of the Wrist, Volume II
Keith Code
Acrobat Books
Venice, CA, 1993

The Unfair Advantage
Mark Donohue and
Paul Van Valkenburgh
Dodd, Mead and Company
New York, NY, 1975 (out of print)

Winning—A Race Driver's Handbook
George Anderson
Motorbooks International
Osceola, WI, 1993

Periodicals

Circle Track
Petersen Publishing Co.
6420 Wilshire Blvd.
Los Angeles, CA 90048-5515
800/800-6825
Technical information and race coverage for Winston Cup and other stock cars with some info on Indy cars.

Indy Car
ICR Publications, Inc.
617 S. 94th St.
Milwaukee, WI 53214
800/432-4639

Race coverage and happenings within Indy car, Indy Lights and Formula Atlantic.
http://www.icr.com

Open Wheel
Open Wheel Publishing, Ltd.
277 Park Ave.
New York, NY 10172
Race coverage and technical information on Sprint Cars and Midgets.

Racecar Engineering
Q Editions Ltd.
33 Banshead Rd.
Chaterham, Surrey CR3 5QG
England

Available in the U.S. from:
Eric Waiter Associates
369 Springfield Ave.
Berkeley Heights, NJ 07922
908/665-7811
High-end technical information on all forms of road and oval racing. Articles written by some of the foremost racing engineers in the world.
http://ewal.com/mghre.html

Racer
Racer Communications, Inc.
1371 E. Warner Ave., Suite E
Tustin, CA 92680
800/999-9718
Racing coverage of the premier racing series.
http://www.racer.com

Sports Car
Published by Pfanner Communications, Inc., but available through:

Sports Car Club of America, Inc.
9033 E. Easter Place
Englewood, CO 80110
303/694-7222
The house magazine of the Sports Car Club of America, Inc. (SCCA). Covers the SCCA and pro racing.

Videos

Circuit	Available from Classic Motorbooks.
Drive to Win	Available from the Jim Russell Drivers School.
Going Faster	Available from the Skip Barber Drivers School.
Grand Prix	Available from Classic Motorbooks.
Le Mans	Available from Classic Motorbooks.
Winning	Available from Classic Motorbooks.

Appendix C

Sources

Aeroquip Corporation Performance Products Division P.O. Box 700 Maumee, OH 43537-0700 http://www.aeroquip.com	AN hose and fittings
Automotive Racing Products 531 Spectrum Circle Oxnard, CA 93030 http://www.arp-bolts.com	High-quality threaded fasteners
ACCO Industries Inc. Cable Controls Group 220 Industrial Drive Milan, TN 38358	Control cables
Baker Bearings 2865 Gundry Ave. Long Beach, CA 90806	Rod ends and related products
BAT Ltd. 1748 Independence Blvd., Unit G-2 Sarasota, FL 34234	Small Formula car equipment
BRITS 28921 Arnold Dr., F-6 Sonoma, CA 95476 707/935-3637	Hard-core racing parts

Coast Fabrication 17712 Metzier Lane Huntington Beach, CA http://www.coastfab.com	AN threaded fasteners
Cablecraft, Inc. 2011 S. Mildred St. Tacoma, WA	Control cables
Dixon Steel Fabrication RD1 Pleasant Mount, PA 18453 570/448-2727	Metal working tools
Donnybrooke Motor Racing Equipment 319 Lake Hazeltine Dr. Chaska, MN 55318	Suits, helmets, mirrors, etc.
Earl's Performance Products 825 E. Sepulveda Carson, CA 90745	AN hose and fittings
The Eastwood Company 580 Lancaster Ave. Malvern, PA 19355 http://www.eastwoodco.com	Metalworking tools
Mark Weber Exclusive Sports Car Photography 430 Foote St. St. Louis, MO 63119 314/961-4571	Auto racing photography
Fast Forward Components 507 Redwood Ave. Sand City, CA 93955	Brake and driveline parts for FF, etc.
Healthy Living, Inc. www.HealthyLiving.com	Nutritional products for racers

Jeff Neil JN Productions P.O. Box 293024 Lewisville, TX 75029 800/479-5636 jnprod@pobox.com	Promotional videos
Metric and Multistandard Components Corp. 2200 Century Drive Irving, TX 75062 214/358-4106 http://www.metricmcc.com	Metric and hard-to-find threaded fasteners
Ron Minor Racing 6511 N. 27th Ave. Phoenix, AZ 602/242-3398	Suits, helmets, mirrors, etc.
David P. Morgan 4213 Knollton Rd. Indianapolis, IN 46228 317/297-8889 dmpco@juno.com	Professinal sponsor hunter
Pegasus Auto Racing Supplies 2475 S. 179th St. New Berlin, WI 53146 414/782-0880 http://www.execpc.com/~pegasus	Hard-core racing parts
Russ Harriott Sports and Entertainment International 41495 Chaparral Drive Temecula, CA 92592 909/693-4144	Professional sponsor hunter
SPS Technologies Aerospace and Industrial Products Division Highland Ave. Jenkinstown, PA 10946 http://www.spstech.com	High-quality threaded fasteners

Taylor Race Engineering
2030 Ave. G, Suite 1102
Plano, Texas 75074
800/922-4327
http://www.racegear.com

Gears and driveline products for small Formula cars

Translantic Racing Services
5730 Chatahoochie Industrial Park
Cumming, GA 30131
800/533-6057
404/889-0499

Hard-core racing parts

Wicks Aircraft Supply
410 Pine St.
Highland, IL 62249
800/221-9425
618/654-7447
http://www.qpg.com/w/wicks/

Composite materials, steel and aluminum, An fittings, etc.

Williams Engines
2701 W. 47th
Shawnee Mission, KS 66205

Small Formula car engine builder

Appendix D

Sanctioning Bodies

ACCUS—Automobile Competition Committee for the United States
1500 Skokie Blvd., Suite 101
Northbrook, IL 60062

ASA—American Speed Association
202 S. Main Street
P.O. Box 350
Pendelton, IN 46064
765/778-8088
http://www.asaracing.com

CART
390 Enterprise Court
Bloomfield Hills, MI 48013

Chicago Historic Races
825 W. Erie St.
Chicago, IL 60622

Grand American Road Racing Association
1801 W. International Speedway Blvd.
Daytona Beach, FL 32114
904/947-6691
www.grand-am.com

HMSA—Historic Motor Sports Association
P.O. Box 30628
Santa Barbara, CA 93130
805/966-9151

NASCAR—National Association for Stock Car Automobile Racing
P.O. Box K
Daytona Beach, FL 32015
http://www.nascar.com

PSR—Professional Sportscar Racing
P.O. Box 10709
Tampa, FL 33679-0709
http://www.professionalsportscar.com/

SCCA—Sports Car Club of America
P.O. Box 3278
Englewood, CO 80112
http://www.scca.org

SVRA—Sportscar Vintage Racing Association
2725 W. 5th North St.
Summerville, SC 29483
http://svra.com

USAC—United States Auto Club
P.O. Box 24001
Indianapolis, IN 46224
http://www.usacracing.com/

Walter Mitty Challenge Vintage Group
P.O. Box 550372
Atlanta, GA 30355-2874

Appendix E

Glossary

Ackerman steering Steering linkage designed to turn the inside wheel at a sharper angle than the outside one.

alligators (or alligator's teeth) Sometimes race track designers or owners decide where racers will drive and where they will not. When large concrete traffic buttons are added to an FIA curb (see below), the resulting structure is something with which errant drivers can (and do) tear out oil pans, tub bottoms, lower portions of frames and expensive gearbox castings. Alligators can also be made of parallel concrete strips with only slightly less damage potential.

AN hose Hose originally designed and produced to meet Air Force/Navy specifications. Now it is available from a number of manufacturers for general use without meeting military specs and without the high cost those specs produce. AN hose is used on a race car for fuel, oil, water, and hydraulic lines.

anti roll bar A torsion bar mounted across a car connecting the wheels at each end in such a way as to resist the rolling motion of the chassis in response to the weight transfer caused by cornering.

apex joint A Mil Spec U-joint assembly used in steering and shift linkages. An apex joint is small, light, precise, and somewhat expensive. It is sometimes call a "helicopter joint."

A/SR A Sports Racer.

backmarker A slower car running at the rear of the pack. Usually used in the context of faster cars approaching, i.e., "He's lapping a backmarker."

blocking A rather ambiguous term which indicates a defensive passing maneuver. If a car continually moves across the track to keep from being passed he is considered to be blocking.

blueprint To modify each part of an engine to bring it up to the maximum limit of the specifications. All stock specification engines (spec engines) should be blueprinted prior to racing.

brake bias The distribution of braking force between the front and rear axles as a percentage. Most cars have some means of changing this distribution either in the paddock or on the track.

B/SR B Sports Racer.

bump Upward movement of a wheel or compression of a shock absorber.

bump steer Toe change of a wheel in response to vertical wheel movement.

camber The angle from vertical of the plane of a wheel. Also used to indicate the asymmetry of a wing as seen in the chord plane.

carbon fiber A synthetic fabric used like fiberglass in the construction of composite chassis, wings, helmets, etc. Although it is somewhat lighter than fiberglass its real advantage is the tremendous stiffness and tensile strength it produces in a finished part.

car control A driver's ability to keep the attitude of his car (yaw angle) and it's dynamic weight distribution correct at all speeds and force levels.

CART Championship Auto Racing Teams.

caster The angle from vertical of the king pin or the angle described by the ball joints of a suspension as seen in side view.

center of gravity Term often used to refer to the center of mass. An imaginary point that is the average of a car's mass in both the front or rear view and the side view.

C_f Coefficient of friction.

CG height The height above the ground of the center of gravity.

chicane An artificial 'S' bend put into a track for the purpose of reducing speeds.

chord When looking at a section of a wing from the side, the length of a straight line from the center of its leading edge to its rear tip.

coil over A spring/shock unit in which the coil spring is mounted coaxially over the shock absorber.

compression The upward movement of a wheel from static position, or the compression of a shock from it's static position. Also used as a measure (in pounds) of the effectiveness of the seal of rings and valves in an engine's cylinder. See also "leakdown" in this context.

contact patch The area of the face of a tire that is in heavy contact with the track surface at any given point in time.

corner (relating to car) The complete suspension, brake, and wheel assembly at one corner.

corner (relating to track) A turn of any angle, from a "kink" to a "hairpin."

corner entry The first part of a turn, from the turn in point to the point of throttle application.

corner entry understeer The understeer that occurs within the area of the corner entry. All cars have some corner entry understeer, but in the best ones it goes away so quickly that it is imperceptible to the driver.

corner exit The area of a turn from hard application of throttle to the track out point.

corner station The headquarters of the area worked by the crew of a particular corner. The corner workers assist drivers of spun or off course cars and show flags from the corner station.

corner weight The weight supported statically by each of the four tires on a car.

crack check Examining parts that are suspected of containing cracks. Several methods can be used, but the most common are the Magnaflux method, which requires machinery not commonly available at racetracks and the dye penetrant method, which uses fluorescent dye. Dye penetrant kits are available at welding supply stores.

C/SR C Sports Racer.

CV joint (constant velocity joint) A joint that, unlike a U-joint, does not result in four acceleration/deceleration periods during a revolution when it is operated at an angle. CV joints are sometimes called "Rezepa joints" after their inventor.

data acquisition The use of an on-board computer to record data sent by sensors mounted on various parts of the car. That collected data is then downloaded into a laptop computer for analysis. Through data acquisition, information about a car's handling and performance can be obtained which is unavailable by any other means.

dash Refers to unit of measure for AN hose and bolts. AN hose and bolts are identified in size by a dash number indicating sixteenths of an inch. A dash 3 is 3/16 and a dash 10 is 5/8 of an inch. Dash has no relation to instrument panel.

dead pedal Attachment to the chassis to the left of the clutch pedal which the driver uses to brace himself when the clutch is not being used.

Denver Used to mean the SCCA administration or headquarters.

differential The mechanism inside the rear end assembly or transaxle which transmits torque to the wheels without solidly coupling the wheels to each other.

DNF Acronym for Did Not Finish.

DNS Acronym for Did Not Start.

double clutch A means of matching gear speed with the clutch pedal to facilitate shifting.

downshift To select a gear one or more gears lower in ratio on approach to a turn. Downshifting is not to be used to slow the car unless the driver is nursing failing brakes.

draft To follow another car closely for reducing aerodynamic drag. See also slipstream.

drift The cornering attitude of a car in which all four wheels are parallel to each other, but at an angle to the car's direction of motion. A driver's ability to drift a car implies very good car control.

droop A wheel or suspension's vertical movement from ride height downward.

droop limiter Any mechanical device that limits the droop a front wheel can achieve. Used to reduce inner rear wheel spin.

dry sump A lubrication system in which the oil is contained in a separate tank and the engine's oil pan is essentially "dry."

dynamic The state of a race car in motion on the track. For example, the dynamic weight of a car using wings is different than its static weight, which is measured in the garage.

dyno Short for dynamometer. A dyno is a machine that measures an engine's torque output. It is also valuable for breaking in a new racing engine and ensuring no leaks or other problems.

F/1 Formula One.

F/2000 Formula 2000.

F/3 Formula Three.

F/3000 Formula 3000.

F/500 Formula 500.

F/5000 Formula 5000.

F/A Formula A or Formula Atlantic, depending on time frame.

false grid Area used to group cars of one class when another class is already on the grid.

F/B Formula B.

F/C Formula C or Formula Continental, depending on time frame.

F/F Formula Ford.

FIA curb A low, gradually sloping curb lining both the inside and outside of a turn at some tracks. Their gradual nature allows even very low and stiffly sprung cars to drive over them with the inside, unloaded wheels.

fifth wheel trailer A trailer using the inclined plate hitch commonly found on 18-wheel trucks. See also gooseneck trailer.

flat spot Describes the condition of a tire. When a tire is locked in braking, or when a car is spinning, a great deal of rubber is worn off of one side resulting in a tire that is no longer round and is "flat spotted." A flat spot is also refers to a sought-after place in the paddock to park the car to measure alignment settings.

flat Refers to foot flat on the floor without lifting as in, "That turn can be taken flat in third."

flat tow Refers to the tow of a car to the pit area. When a car spins or stalls on the track or otherwise cannot run, but can roll and be steered, it is towed behind one of the track's tow vehicles to the pit area.

FOCA Formula One Constructors Association.

Formula car An open-wheel single-seat racing car. If taken literally, any race car could be considered a Formula car since each class is governed by a set of rules or formula.

F/P F Production.

F/V Formula V.

GCR SCCA's designation for their General Competition Rules, which is one of their rule books. See also specification book.

G force A force equal to that of gravity. Usually used to indicate the acceleration forces acting on car and driver in lateral and longitudinal directions.

gearbox Racer jargon for transmission. Sometimes shortened further to "box" as in, "The new ratios are installed in the box."

getting a tow See drafting and slipstreaming.

gooseneck trailer A trailer built to use a ball hitch inside a pickup bed, not on the bumper.

gray, in the See marbles.

grid The area where cars are grouped before they enter the track for the next session or race. Also used to mean the cars themselves prior to the green flag as in, "The grid is rounding the last turn."

ground effect The aerodynamic effect of the ground in close proximity to an aerodynamic device. Ground effect cars use the composite surfaces of their undersides to reduce the local air pressure in the same way as the underside of a wing does. When close to the track surface, this low pressure is not as easily diluted by ambient airflow.

GTP Grand Touring Prototype.

HANS Safety harness manufactured by Jim Downing.

hairpin Technically, a turn of more than 180 degrees. In practice any U-shaped turn on a track is considered a hairpin. Sometimes also called a buttonhook or switchback.

HDL High density lipoproteins.

Heim joint Heim is a brand name of spherical bearing.

homologation The process by which a make and model of car is approved for competition by the sanctioning body. Usually used for purpose built competition cars.

horsepower Technically, the relationship between torque and engine speed. In practical terms, a measure (together with aerodynamic drag and frictional losses) of a car's top speed potential. Although horsepower numbers are thrown around in bench racing sessions—like frisbees in a park—torque is more meaningful to lap times.

impound Area where, at the conclusion of a race, the top finishing cars are grouped and are subject to inspection for rules compliance. The cars are not accessible to the crew or driver until released from impound.

IMSA International Motor Sports Association.

independent suspension Technically, any suspension type in which the only connection between the wheels on one axle is an anti-roll bar. By this definition, swing axle and McPherson strut suspensions are independents. In the real world a double A-arm suspension is considered an independent.

IRL Indy Racing League.

I/T Improved Touring.

Kevlar A synthetic fabric used like fiberglass in the construction of body panels, helmets, etc. Although not a great deal lighter than fiberglass, Kevlar is much stronger and resists tearing and shearing.

kink A turn on a track of very small angle. Kinks are often found in straights and can be very tough because of the extreme high speeds at which they must be taken.

laminar flow Airflow over a surface which follows the contour of the surface smoothly.

lap time The time it takes for a car to make one lap. The sanctioning body's timing and scoring people record lap times and each team should record their car's times, too. The times will not be identical unless they are taken at the same point on the track.

LDL Low density lipoproteins.

leakdown A test performed on the cylinders of an engine to determine condition. By introducing compressed air into a cylinder and measuring how much leaks out, the condition of valves and rings can be known.

lift The partial or complete closing of the throttle. Also is a measure of how far a camshaft opens the valves.

locker A Detroit locker, a type of limited slip differential.

lock To use the brakes too heavily—as in "locking a tire." May also be caused by a combination of too much brake pedal and incorrect brake bias.

long Refers to gear ratios that are too high, as in, "The gears are too long."

loose When applied to a car's handling, refers to oversteering.

McPherson strut An independent suspension using a lower A-arm swinging vertically to actuate a coil over spring/shock unit attached rigidly to the upright. Common on road going cars like BMWs and many Hondas.

marbles Describes a condition at pro races where, because the rubber compound of the tires is very soft, rubber particles are ground off of the tires in the corners and thrown to the outside of the turn. Driving in this area is exciting at best and can produce expensive results. Marbles can also occur at amateur races even though the rubber is somewhat harder because of the number of cars run during a day. At tracks that are not run frequently, loose asphalt, rocks, gravel, and other debris can cause the same situation.

meatball A flag peculiar to SCCA with a red or orange circle centered on a black background signifying a mechanical problem with a car.

mid part The area of a turn between the point of light throttle application and the point of heavy throttle application with hard acceleration.

Mil Spec Military Specification.

mistake Any action that causes a lap to be less than perfect. It is doubtful that any driver has ever made even one perfect lap. Small mistakes are normal and include missing an apex or a turn in point by a few inches. A larger mistake would be dropping two wheels off the track. Still larger mistakes end in crashes. Among competitive drivers, the one who makes the fewest mistakes in a race many times wins.

monocoque Literally, single skin. A race car frame made from aluminum panels bent and riveted together. See tub. Technically, carbon fiber cars can also be called monocoque, but convention has termed them composite.

MS Mil Spec.

NACA duct A duct for importing air into a race car, shaped to reduce drag as much as possible. An NACA duct is an internal duct with no surfaces protruding outside of the normal contour of the body. It is quite effective on the front half of a car where the airflow over the surface is laminar.

NAS Naval Air Standard.

NASCAR National Association of Stock Car Racing.

NHRA National Hot Rod Association.

normal force An engineering term denoting a force applied at a right angle to the resisting plane.

on the hook Describes a race car that is being towed after an accident. When a race car is involved in an accident that results in inability to roll, it must be brought in by a tow truck by picking it up by the roll bar or cage. This is what is called being "on the hook."

open differential A type of differential that allows the wheels of one axle to spin freely relative to each other.

overrev To run an engine at an RPM above its maximum recommended limit.

PCS Production Car Specifications.

polar moment of inertia An engineering term that describes the degree of concentration of mass about a center point. A barbell with weights at each end has a higher polar moment of inertia than the same bar with the weights moved to the center.

powershift To make an upshift without lifting off of the throttle.

power to weight ratio The weight of a car divided by the horsepower produced by the engine. The muscle cars of the late sixties were about 10 lbs/hp, which is pretty good. Sprint cars are about 2 lbs/hp. Pounds per foot pound of torque would be a more accurate description of a car's accelerative capability.

PSR Professional Sports Car Racing.

push Another term for understeer. The opposite of loose.

rebound The downward movement of a wheel away from the chassis or the extension of a shock absorber.

red line The maximum recommended RPM of an engine.

rev limiter An electrical device that limits the maximum engine RPM that can be attained by interrupting ignition function. No rev limiter will limit excessive RPM due to mechanical forces such as downshifts.

Reynolds number An aerodynamic term relating air speed, density, viscosity, and the length of the surface over which it has traveled.

ride height Another term for ground clearance, used in chassis setup.

rod end A spherical bearing that has freedom in two planes, usually used on suspension connections.

roll center The point about which one end of the car rolls in response to lateral weight transfer in a turn. In engineering terms, the roll center is the point of force resolution when cornering forces are applied to suspension parts.

roll steer Describes the condition of a car in which toe change caused by chassis roll has affected both wheels on an axle in opposite directions, i.e., one goes to toe in and the other to toe out. Under these circumstances, the car is said to have roll steer.

roller Technically, a car without an engine. Many modern mid-engined cars use the engine as a stressed or semistressed member, and will not roll without the engine in place.

Runoffs™ SCCA's term for their amateur National Championship races.

S/2 Sports Racing 2000.

SAE Society of Automotive Engineers.

SCCA Sports Car Club of America.

segment time Refers to the time it takes to go a specific distance. When testing, a team can learn where a car is fast or slow by measuring the time it takes go a specific distance at a certain place on the track. Segment times are usually taken from the entrance to exit of certain corners, or from a corner exit to the entrance of the next one.

short Describes gear ratios in the transmission or rear end that are too low, as in, "He's geared too short."

short shift To upshift below the normal shifting RPM.

shunt A term used mostly by the British usually meaning minor crash. Shunt can also mean a heavy crash as in, "That was a bad shunt."

slide An oversteering attitude of a car. See also drift.

slip angle The difference in angle between the rolling direction of a tire and the direction of the rubber in the tread of the tire where it is in contact with the track surface.

slipstream To follow a car closely for the purpose of reducing aerodynamic drag. See draft.

SOHC Single overhead cam.

SOM Steward of Meet.

specification book A spec book for a certain category of racing. SCCA has spec books for each category of racing which include the details of SCCA rules for each class within that category.

stagger The difference in circumference of a pair of tires on an axle. Usually used in chassis setup of oval track cars.

straight Short for straightaway. The portion of a track where full throttle can be used for a long period. Sometimes straights include turns that are taken flat out, such as the back straight at Road Atlanta or the curved front straight at Long Beach.

swing axle The type of independent suspension most commonly found at the rear of early VW Bugs (however, Mercedes built a "low pivot swing axle" as early as the 30s). A swing axle usually results in high camber change and a high roll center location.

tall Describes gear ratios in the transmission or rear end that are too high, as in, "He's geared too tall."

tearoff Thin transparent plastic sheets that attach to the visor of a helmet. Several can be attached and torn off individually by the driver as they become dirty.

tell-tale A tachometer function that records the highest RPM reached since it was last reset.

tiger's teeth See alligators.

TIG welder Tungsten Inert Gas welder (heliarc).

tire temp The temperature of the tread rubber of a tire. When tire temp is measured at three points across the face of a tire, a lot of information can be gained about chassis setup.

toe and heel The process of using the right foot on the brake and throttle pedals simultaneously to bring the engine RPM to the proper point to facilitate shifting while braking.

torque A rotational force. Although many of the components of a race car may be discussed in the context of torque, this term is usually used in relation to the torque output of the engine. The torque of the engine is the deciding factor in a car's ability to accelerate out of a turn.

tow money Money paid to racers to help offset travel expenses. Many pro races and some amateur events (e.g., SCCA's Runoffs™) pay money to racers who tow a great distance to get to the event, which helps to offset their travel expenses.

trail braking Buzzword for continuing to brake lightly after the steering wheel has been turned into a corner. See Chapter 6.

track width The lateral distance between the wheel centers at one end of the car. Occasionally used to mean the width of the track surface.

trailing throttle oversteer Refers to the oversteer that results when a driver lifts off of the throttle in a turn and the contact patches of the rear tires are unloaded, reducing traction.

tub The frame of a monocoque car.

tube frame The frame of a car built from tubing, usually steel.

track out point The maximum lateral point on the track to which the car is allowed to drift out when exiting a turn.

turn in point The point at which the wheel is turned to initiate a change of direction when approaching a turn.

unibody A monocoque frame built of steel panels by a major car manufacturer. Most current road going cars are built by this method.

upright On a purpose built racing car, the part that holds the axle between the A-arms. On production based cars the upright is usually called the spindle.

upshift The selection of a higher gear when accelerating.

USAC United States Auto Club.

VDS Cars build in the '80s by Count Van Der Straten.

weight jacking Many oval track cars have adjusters on the chassis springs at each corner enabling the crew to make quick adjustments to the corner weights. See corner weights.

WSC World Sports Car.

Zyglo Brand name for a type of dye penetrant crack checking equipment.

Epilogue

The Road to Success

Effort equals success.—Roger Penske

Although we can easily understand Roger's message, I believe that more than just effort is necessary to ensure success. The essential ingredients include crew members who are intelligent, experienced, and dedicated. As in business, the people involved are the most important part of the equation. They steer the team in the direction that seems most appropriate. The more intelligent, experienced, and dedicated they are, the more accurate their steering is likely to be. Success is also directly linked to the resources available. Often racers can be heard bemoaning the fact that another team is better financed and therefore takes home all the marbles. During May 1995, when both of their drivers failed to qualify for Indy, the Penske organization showed us very painfully that it does not matter if a team has seemingly unlimited resources. (This was after winning that race and the championship the previous year.) Races are won each weekend by teams that are underfinanced because those teams are able to creatively solve the day's problems better than their competition. Having the money to purchase all of the bells and whistles is a noble goal, and is one toward which every team should work diligently, but it is not a prerequisite for success. Obviously, some money is required, so money *is* a part of the equation. Once the people and the money are in place, *then* effort is the catalyst that makes them work. I will take the liberty of quantifying Mr. Penske's observation by stating:

$$\text{Success} = (\text{Quality People}^2 \times \text{Effort}) + \text{Money}$$

However, due to the dynamic nature of racing, even this formula must be often revised, and it is likely that next year or the year after will present slightly different challenges to the racer. Ron Denis observes, "The key to continuous success is to realize that the recipe for success is constantly changing."

At the beginning of almost every racing career is a person who buys a car and goes out to race. In time, he usually becomes the focal point of the team. Throughout this book, we have discussed not only the skills he should develop to become successful, but also the attitudes with which he must view his endeavor. The attitudes characteristic of a racing driver include humility, which enables him to continue to be a student throughout his career, continually learning new techniques, and technology, which can help make him faster. Also important is an analytical approach to preparing and driving a race car and an effort to constantly know what is *really* going on as opposed to what we only think is happening. These attitudes even include integrity. Racing is a small community, and a racer who does not deal honestly with all of the other people involved will soon be branded as an ego racer and boycotted by those who are both serious and honest. The annals of the sport are littered with those who came in like a flash, took advantage of all those they encountered, and beat a hasty retreat into another hobby. Due to the egos involved in racing, enemies are much too easy to make, even when a racer does live by high ideals. Endurance is also required. Racing is expensive in terms of the money, time, and the effort required to win. In any given race, there are 25 or 30 who lose and only one winner. Races are lost far more often than they are won, and it takes great endurance to persevere under these conditions.

Two of the most important attitudes a driver must develop to be successful, though, are confidence and determination. It is normal to have "butterflies" while sitting on the grid waiting for a race to begin. Every driver experiences this feeling prior to every race. However, deep within his psychological makeup, underlying his nervousness must be a sense of calm that comes from the confidence of knowing that he can do the job. Racing is a difficult sport, but it is one at which anyone with the basic physical traits can excel. It is true that the Jim Clarks, Giles Villeneuves, and Ayrton Sennas of the world are extremely rare. There are thousands more drivers, though, who may not have a natural gift, but are successful nonetheless. They all share the one quality that is essential, the belief that they can do it. The truth is that each of us can do anything we want as long as we want to do it badly enough. It is absolutely critical for a driver to have this conviction at the very core of his being.

The desire to succeed spawns the determination needed to win. It is easy to spot the drivers who have learned the driving, racing, and setup skills that are required to move up through the pack. The real standouts, though, are those who just plain drive the wheels off the car to move to the front. They may not have the best car, but they run with the guys who do because they are *determined* to win. This same determination shows in their off-track behavior, too. A driver who has this kind of determination will learn chassis setup if that is not his strong point, he will find the best car if he does not already have it, and he will find the sponsors to provide the funding to make all the rest happen. The determination that a driver must have in order to get to the top cannot be overemphasized. Racing is such a complex activity and so demanding in nature that, if a driver is not absolutely determined to make it happen, the obstacles that inevitably get in his way will eventually stop him. If the desire is strong enough, he will find a way to make it happen.

In racing, as in practically every other field of modern endeavor, success is a term that does, and should, mean different things to different people. To some, it may mean winning Indy. Others may feel they are successful if they are competitive and have a chance to win each club race they run. Your success must be defined by your goals. One prerequisite to being successful, then, is to actually *have* goals. As you gain experience, those goals, and your definition of success, are likely to change.

For many amateur racers, family and career responsibilities keep goals moderate. If your goals do not include winning Indy, it may be tempting to think that you can go through the motions of being a racer and achieve rewarding results. It has been my experience that this is not the case. At every level, racing demands a systematic, analytical approach. Each aspect of the sport must be carefully planned and every detail performed correctly. Haphazard car preparation, improvisational lines through the turns, and lazy attempts at obtaining sponsors will not get the job done. Those drivers who constantly run in the middle of the pack, have mechanical problems, don't drive hard enough or throw the car at a turn, burn up tires, and come out of it slow, are almost always frustrated and wonder why they have so many problems. Few things in racing make a racer feel worse than the long tow back home after an event at which the header fell off, the engine did not perform correctly, the car had to be rushed through tech because it did not receive an annual tech at home, two members of the crew arrived at the track late, and a widget needed to repair the car was left on the bench in the garage. This scenario sounds like an outrageous exaggeration, but it happens all across the country each weekend. Part of every racer's goal must be to reduce this frustration level, and the way to do this is through organization.

Things don't go smoothly at a race by accident. When crew members are polishing the car prior to a session and the driver takes the car out and turns the fastest time, it happens because the team planned it that way. This planning includes the work back at the shop, ensuring that the proper parts arrived on time and that the car was ready when loaded in the trailer. It also includes ensuring that the crew was fresh and in good spirits, the driver knew the track, the driver and engineer knew how to set up the car, and the wives and girlfriends had travel accommodations, meals, and kind words for everyone involved. This may sound like a minor point, but something as simple as a screwed-up hotel reservation can have a marked effect on the crew's attitude, and their performance can affect the performance of the car. No detail can be overlooked; no detail is unimportant. The Andrettis and Unsers may race for a living, but they would find other vocations if racing were not fun. Many others race *only* because it is fun. If the crew/driver/wife or anyone else involved with the team is not happy, no one will have fun. Plan your events carefully, from start to finish, and they will run smoothly, everyone will have more fun, and you will have a better chance of accomplishing the goal all racers have in common—winning races.

Index

Abbreviations are used to indicate figures (*f*), tables (*t*), and photos (*p*).

About the Author

The author (left) and one of the drivers that he works with, Jeff "Hound Dog" Harrison.

Robert Metcalf is a racing engineer who consults with racers worldwide. He has worked with dozens of drivers, from beginners to some of the top professionals, and with many different types of cars, including Formula Cars, Production Cars, GT Cars, Stock Cars, Unlimited Pike's Peak Cars, Can-Am Cars, Long Distance Prototypes, and virtually every other type of race car that runs on dirt or pavement and turns corners. His drivers have scored an impressive three National Championships and over a hundred race wins. One was even awarded SCCA's Kimberly Cup for most improvement in a season.

Metcalf is also a race car driver. He began racing Karts in 1958, when both he and Karting were young. He raced Karts until he was old enough to drive Formula Fords in SCCA and continued through Formula 5000 and Sprint Cars. In 1980, he realized that he had become a better racing engineer than driver and hung up his helmet so he could concentrate on setting up cars for other drivers. He still retrieves his helmet from time to time for an occasional test or race.

Metcalf makes his home in Dallas, Texas. He is a frequent contributor of technical articles to racing publications and has presented numerous seminars on vehicle dynamics. Outside of racing, his interests are sailing and blues guitar.

For more information, see RACECARCONSULTANTS.com.